Space Science

Space Science

Edited by **Jean Tabor**

New York

Published by Callisto Reference,
106 Park Avenue, Suite 200,
New York, NY 10016, USA
www.callistoreference.com

Space Science
Edited by Jean Tabor

International Standard Book Number: 978-1-63239-573-3 (Hardback)

Printed in the United States of America.

Contents

Preface

Over the recent decade, advancements and applications have progressed exponentially. This has led to the increased interest in this field and projects are being conducted to enhance knowledge. The main objective of this book is to present some of the critical challenges and provide insights into possible solutions. This book will answer the varied questions that arise in the field and also provide an increased scope for furthering studies.

This book consists of state-of-the-art information regarding the extensive field of Space Science. Space Science is a vast term which describes the distinct fields of research in science related to the study of the Universe, usually excluding Earth's atmosphere. These include physics and astronomy, aerospace engineering, spacecraft technologies, advanced computing and radio communication systems. This book is a distinct publication on space science, providing significant information compiled through a scientifically rigorous process. The book has been strategically structured to take the reader on a fascinating journey beginning from the very surface of earth and extending up to the periphery of this universe. It provides an updated review on space exploration along with analyzing the role of upcoming nations. An overview on Earth's initial evolution during its ancient ice age has also been provided in the book. Towards the end, a revaluation of some of the aspects of planetary dynamics and satellites has also been provided, followed by interesting discourses on new advances in cosmology and physics of cosmic microwave background radiation.

I hope that this book, with its visionary approach, will be a valuable addition and will promote interest among readers. Each of the authors has provided their extraordinary competence in their specific fields by providing different perspectives as they come from diverse nations and regions. I thank them for their contributions.

Editor

Part 1

Space Exploration

How Newcomers Will Participate in Space Exploration

Ugur Murat Leloglu and Barış Gençay
TUBITAK Space Technologies Research Institute
Turkey

1. Introduction

Space Exploration, one of the hardest achieved successes of mankind, is defined as all activities geared towards exploration of outer space using either space technology or observations from Earth, though sometimes the latter is not considered as part of space exploration (Logsdon; 2008). In this chapter, we will exclude observations from the Earth or the low Earth orbit (LEO) and scientific LEO missions that explore plasma sphere, which deserve dedicated study, especially because the opportunities cubesats offer to newcomers who want to contribute to space science (Woellert et al., 2010). While exploring our planet from its core to the surface and beyond, space studies has provided good leverage for science, technology and spin-off applications. Since the beginning of the space age, whose onset is generally accepted as the year 1957 when Sputnik-1 was launched, our knowledge about outer space has increased at an accelerating pace, an achievement made possible by developments in space technology. Mankind has succeeded to send satellites, landers, and rovers to other planets and their satellites, built an orbiting space station, analysed samples of other planets' soil, atmosphere and magnetosphere, performed regular launches to various earth orbits, planned regular touristic rides to space and even sent men to the Moon.

The Treaty on Principles Governing the Activities of States in the Exploration and Use of Outer Space, including the Moon and Other Celestial Bodies, otherwise known as the Outer Space Treaty in short, defines basic principles for use of space. Although the Outer Space Treaty states that "The exploration and use of outer space, including the Moon and other celestial bodies, shall be carried out for the benefit and in the interests of all countries" and has been signed by the majority of the world's nations, as shown in Figure 1, until recently space exploration has actually been a privilege for only the few developed countries who could actually 'touch' the space.

Nevertheless, the number of countries who have initiated space programmes to benefit from space is increasing. Several large countries, like India and China, were early to establish their space programs and have been followed by many others. Although the initial steps are generally small and focus on immediate needs, the programmes eventually involve more scientific content; enabling new nations begin to contribute to the exploration of space, a trend that can be called "democratization of space"[1]. These newcomers, mostly from newly industrialized countries, and Asian nations in particular, are paving the way for intensive space exploration activities.

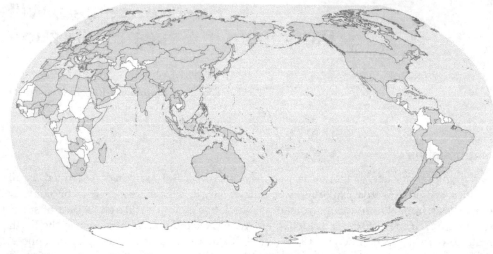

Fig. 1. Outer Space Treaty Signatory States. Blue: signed and ratified, green: signed only.

In this chapter, after a short summary of space exploration, we first try to draw a picture of the democratization of space, i.e. the joining of more nations to the space club. Then, with a focus on space exploration, we discuss possible opportunities and advantages as well as difficulties for the newcomers.

2. A short summary of space exploration

From the very beginning, humanity's desire to reach celestial objects was reflected in the mythologies of various civilizations. An example from Turkic mythology is the celestial journey of the Shaman to the fifth level of the Sky, Polaris, after sacrificing a white horse. According to their belief, the Moon was on the sixth level where humans could not reach (Gömeç; 1998). However, with Kepler's laws describing the movement of planets around the Sun and following breathtaking scientific and technological developments, travel to the Moon became a reality during the space age.

Following the end of WWII, the Soviet Union shocked much of the world with its launching of the Sputnik-1 satellite that transmitted periodic pulses and Sputnik-2 satellite carrying a dog as a passenger, launched onboard the modified Russian R-7 ballistic missiles on October 4, 1957 and November 3, 1957, respectively. The US responded immediately with its own launch of the Explorer-1 satellite on January 31, 1958.

---

[1] The term is generally used to refer the right of individuals to reach orbit in the context of space tourism, however, here we use it for the right of nations to reach orbit and benefit from space in the collective sense.

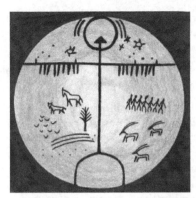

Fig. 2. A drawing of a shaman's drum depicting the conceptualization of the universe by ancient people. Upper part is the sky. (from Wikimedia, originally from (Anokhin, 1924))

The ensuing Cold War between the US and Soviet Union fuelled a fierce race to achieve tactical and strategic space superiority. Space technology developed from, sending first animals then robotic rovers, probes to the Moon, Mercury, Venus and Mars then finally humans to different targets in space including the Earth's orbit and the Moon. In this short period of about 50 years, even the frontiers of our own solar system were explored after Galileo; plans to send probes beyond Jupiter became part of everyday life and space proved to be an economic, diplomatic and strategic tool for those participating the race. Missile race in the 40's resulted in the Moon race in the 50's, followed by the deep space race in the 60's. The countries who pioneered the space race not only led space exploration but also benefited from the return on investment through the technological spin-offs that later achieved commercial success; and industrial mechanisms that turned into political power. Although these scientific, technical, and financial rewards improved mainly the lives of their own citizens, the increase in the base of knowledge, advances in productive capabilities, expansion of the range of economic activities, and enhancements of geopolitical positioning also served to inspire all of mankind.

2.1 Moon rush

Following the successes of the Sputniks and Explorer-1, the Russian Luna-1 satellite became the first spacecraft to escape Earth's orbit in January 2, 1959. On September 12, 1959, the Russian Luna-2 was launched and impacted on the Moon's surface two days after the launch. Luna-3 was launched in October 4, 1959 and became the first manmade object to reach and photograph the dark side of the Moon while the American Pioneer 1, 2, 3, 5 and 6 satellites failed during their launch towards the Moon. In April 12, 1961, Soviet Russia made an enormous step in the history of space exploration when cosmonaut Yuri Gagarin became the first man to successfully orbit the Earth. Shortly thereafter, American astronaut Alan Shepard completed the first suborbital flight in May 1961. On February 3, 1966, the Russian Luna-9 satellite completed a soft landing on the lunar surface. In the same year, the Luna-10, 11, 12 and 13 successfully reached the Moon orbit and the Luna-13 landed on the Moon's surface.

Between 1966 and 1968, unmanned Apollo-1, 2, 3, 4, 5 and 6 spacecrafts were launched on board of Saturn-1B and Saturn-5 launch vehicles. Manned missions of Apollo started with

Apollo-7 in 1968, which carried a crew of three into the Earth orbit. It was followed by Apollo-8 with a crew of three that completed the world's first manned mission around the Moon between December 21 and 27.

Meanwhile, Soviet Russia was developing "Zond" and a prototype of the spacecraft for manned circumlunar flight flew around the Moon.

Between July 16 and 24, 1969, The astronauts of Apollo-11 landed and walked on the surface of the Moon. They became the first men to walk on a celestial object other than the Earth. After the success of Apollo-11, the USA successfully completed five more Moon expeditions with the Apollo series. After several Russian launch vehicles and sample-return spacecrafts failed to reach the Moon, the Soviet government officially cancelled the N1-L3 program in 1976. Until that date, Russian rovers Lunokhod-1 and 2 landed on the Moon safely in 1970 and 1973, respectively. Following these events, moon rush, turned into deep space rush starting from Mars and Venus.

2.2 Deep space rush

In light of the experiences and developments of the Moon race, Soviet Russia and the United States considered exploration of the inner planets as well. Unsuccessful Mars and Venus probes were launched towards their destinations in the early 1960's. After many disappointments and very expensive trials, for the first time in the history, the American Mariner-4 satellite transmitted 21 images and bountiful scientific data in 1964 at a distance of approximately 10,000 km from Mars. US Mars-1 to Mars-7 and Russian Phobos-1 & 2 were also sent to Mars until 1988, and some of them returned valuable data. The Russians responded with the Venera-3 satellite. The lander penetrated the atmosphere of Venus in 1966, another first time event. Following Venera-3's success, a series of landers, Venera-5 to - 16 were sent to Venus until 1983. Exploration of Mars and Venus continues even today, with the US sending two Pioneers to Venus; six more Mariners, two Vikings to the Mars and two Voyagers, three Pioneers to Jupiter until 1978 and many more up to day. Although the space race may seem to have slowed down after the 1980's due to financial reasons, it is unlikely that it will ever end as human interest in space continues to this day with major projects such as the International Space Station.

The technology derived from the Sputnik missions has led to key developments in modern communication, earth observation, meteorology, early warning and scientific satellites that have improved and become the part of everyday life on Earth.

While Soviet Russia and the US were continuously conquering the outer space, the rest of the world seemed reluctant to proceed on the same way and did not join the race for some time. Recently, the investments made by the Asian states in lunar programmes have also increased global interest in the Moon. It is likely that the US, Russia and the European Union will also make significant investments in this direction soon.

2.3 Europe

In 1964, the European Space Research Organization (ESRO) was founded by 10 European nations with the intention of jointly pursuing scientific research in space. ESRO was merged with ELDO (European Launcher Development Organisation) in 1975 to form the

European Space Agency (ESA). In 1970's when Russia and America were flying to Mars and Venus, Europe had just formed its organization devoted to scientific, civilian space applications.

Europeans started developing sounding rockets in 1964, which were followed by several scientific satellite projects in 1968 called ESRO I and ESRO II. After that, Highly Eccentric Orbit Satellite (HEOS) for measurements of plasma, magnetic field and cosmic ray particles and the Thor Delta program for stellar astronomy were started.

The race to the Moon, Venus, Mars and comets did not generate the same interest in Europe as among Russians and Americans during the height of the Cold War. Rather, Europe's long term projects mainly focused on remote sensing, space science, the International Space Station (ISS) and telecommunication. ESA's only mission to the Moon was launched in 2006, 31 years after its establishment.

2.4 Japan

Japan is the first country in Asia to follow the developments in the rest of the world, founding the Institute of Space and Astronautical Science (ISAS) in 1950. The first satellite was launched in 1970 with the indigenous L-4S rocket. In the beginning of its developmental phase, the National Space Development Agency of Japan (NASDA) used a US license to produce rocket engines, which paved the way to the first launch vehicle developed in Japan, the H-II, which was launched in 1994.

The first Japanese missions beyond Earth orbit were launched in 1985 to observe the Halley comet with two observation satellites. The missions were performed together with the Russian and European Space Agencies as part of a joint space exploration program. Japan is also the first Asian country to launch a lunar probe with a satellite called Hiten in 1990. They even sent an orbiter to the Sun in 1991 and sponsored an astronaut mission as part of US Shuttle program in 1992. The first Japanese interplanetary mission, the Mars Orbiter Nozomi (Planet-B), was launched in 1998.

Briefly, Japan invested heavily in exploration of space and space science; astronomy, technology tests, lunar explorations, solar sail research and even sent probes to asteroids and the Moon.

In April 2005, Japan announced ambitious new plans for a manned space programme, including landing on the Moon by 2025. The country now wants to have human presence in space along with unmanned scientific planetary missions and also has ambitions to open a permanent base on the Moon and manned spaceflights around the year 2020.

The Japanese ride to space is supported by their ability to access to space by means of their own indigenous launch vehicle, just like Russia and the US. Sufficient financial support from the government and moral support from society also stimulate Japan's special interest in space. However, it is uncertain today if Japan will continue to invest at the same pace, due to the devastating impact of the Sendai Earthquake and Tsunami in 2011, whose estimated cost is around several hundred billion US $. Probably, some of the funds, which were allocated for space projects, like other government spending, will be transferred to the recovery of earthquake devastated areas and export oriented Japanese economy.

3. Democratization of space

In the early stage of the space age, almost all space activities were carried out by a small number of developed countries and the USSR. However, an important development in recent years is, as we call it, the democratization of space. Increasingly, nations who want to exploit space for the good of their citizens and to boost national development have stepped into the space technology arena. Some large countries had already initiated their space programs as early as 1950's. China and India comprise the category of newly industrialized countries that represent 37% percent of world population and have made great achievements in the meantime. As of today, these countries have managed to put their own launch vehicles on serial production and even reached the lunar orbit.

Newly industrialized countries like Brazil, South Africa, Turkey, Thailand, Malaysia and some other nations have taken their first steps mostly through relatively low-cost small satellite technology transfer programs and/or by collaborating with nations strong in space technology. For these nations, most of whom are either in the newly industrialized or developing country category, the first priority is generally satisfying immediate needs and achieve a return on investments as soon as possible. The main focus is generally on earth observation, which is an important tool to support development. Countries who can afford to have also invested in telecommunication systems and launch vehicles.

In parallel to space technology investments for immediate needs, efforts in the domain of space science and space exploration have increased as well. China and India have progressed similarly and initiated their space activities in 1950's. Although they have boosted their activities much later than Russia, the US and Europe, since the 1980's, they have become part of the elite club that is paving the way for the future of space exploration to expand scientific knowledge, develop their country's technical capabilities, and provide employment opportunities for valuable human resources in the areas of space technologies and science. With increasing interest in space, more countries are aspiring and will aspire to space exploration activities by the use of space technology following these examples. After summarizing the space programs of China and India, we will review developments in the rest of the world.

3.1 China

Actually, the technological roots of Chinese space studies can be traced back to the late 1950's. As the space race between the two superpowers reached its peak within the context of the Moon race, China did not want to be left behind and initiated its manned space program in 1971. The first manned program was cancelled in 1972 due to financial reasons. The second manned program was launched in 1992 and led to the successful orbital flight of Shenzhou-5 in 2003. Following this flight, China managed to send men into orbit and successfully bring them back to Earth in 2008, thereby becoming the third nation in the world to accomplish that complicated mission. This success encouraged China to make an official declaration about plans for a manned space station and the Chinese Lunar Exploration Program (CLEP).

Current indications are that China will proceed at its own pace; it was officially announced that participation in the ISS is not on the agenda. To achieve successful orbital operations of a Chinese space station, several expensive and slow steps have to be taken, including construction of dockable space station, extra-vehicular activity trials with space suits, biological, medical, chemical, electronic and electro-mechanical experiments in orbit, and creating a sustainable habitat for the visitors, just like in the ISS.

While the future of ISS is clouded by financial considerations and very small global public interest, China in contrast has expressed self confidence, self reliance, strong determination and future plans for a space station. However, China will most likely conduct fewer and more limited trial missions, unlike the National Aeronautics and Space Administration (NASA) and Russian space agency (Roscosmos) did in the past, to cut costs. Recently in 2011, the world's largest launch vehicle construction facility opened in China and one of the products will be the Chang Zheng-5 heavy lift launch vehicle, which is supposed to be capable of delivering 25 tons to low earth orbit (LEO) beginning of 2014. Once heavy lift capability is achieved, space transportation for landers, rovers and travel of Taikonauts (Chinese version of the term astronaut) from the space station to the Moon, Mars and beyond is theoretically achievable with sustained cash flow.

The outcome of this investment in space will be very useful in many different areas, such as financial, moral and especially political and geopolitical positioning for China.

The start-up of the Chinese space exploration program is Chinese Lunar Exploration Program (CLEP). The Chang'e program is part of CLEP and currently consists of two orbiter spacecrafts that were launched in October 2007 and October 2010, respectively. These satellites have provided data about possible future landing sites and mapped the surface of the Moon. Although key elements of the first satellite were mainly developed and funded by China, international support came from ESA by providing the necessary deep space network for Chang'e missions in return for Chang'e-1 data. Due to the political reasons, China could not benefit from the US Deep Space Network distributed all over the world, which would have enabled continuous communication with spacecraft and accomplishment of Telemetry, Tracking and Control (TT&C) tasks. Thus, the only option for CLEP was to rely on ESA's network. Meanwhile, China upgraded its own TT&C network, which was originally designed for manned space missions, and managed Chang'e-2 mission without any foreign support, thereby achieving independence. Presumably, we hope to assume China will share its valuable sources through regional, international or bi-lateral cooperation with other nations for space exploration.

For the Mars program, China cooperated with the Russian Federation; however, the Russian partners couldn't perform the launch in 2009 when Mars was relatively close to the Earth, so the most favourable launch window was missed due to the delay in the Phobos-Grunt project. This opportunity could have been evaluated as one of the best joint interplanetary outer space explorations had it succeeded. However, the willingness and close cooperation between these two giant states is an emerging and encouraging opportunity for the others, especially those who want to participate in outer space exploration and share the cost of development and launch.

3.2 India

Following the successful launch of Sputnik-1 in 1957, the Indian National Committee for Space Research was founded in 1962, later evolving into the Indian Space Research Organization (ISRO) in 1969.

Following the same development patterns of Japan and China, India invested in earth observation, communication, meteorology, scientific and outer space exploration programs (e.g., the Moon) and formulated its own launch vehicle program to guarantee the access to Earth's orbit. While the space program was formulated with little foreign consultancy and support, the lunar program, Chandrayaan, was supported by international institutions and several countries. Chandrayaan-1 was launched in 2008 and is one of the best and so far, one of the most successful international outer space exploration programs, even though the mission ended earlier than expected. Bulgaria, the United Kingdom, Sweden, Canada, ESA and the United States participated in the mission, contributing various payloads and flew onboard the spacecraft free of charge. Recently, ISRO allocated funds for Chandrayaan-2 & 3 that includes lunar lander and rover segments. Although NASA and ESA would like to participate this project as well, Chandrayaan-2 will mainly be performed with Russian support and estimated launch dates are after 2013 and 2015, respectively. Another key aspect of Chandrayaan-1 is that 11 different instruments, designed by different organizations, worked well and with each other on a single satellite platform. This represents a tremendous achievement in terms of gathering different organizations and technologies under the same umbrella on board a single satellite platform and enabled them to benefit from the same technological standards on a totally non-commercial space mission.

According to Indian officials, the main drive behind the lunar exploration program is to expand scientific knowledge, develop the country's technical capabilities, and provide working opportunities for valuable human resources in the areas of space technologies and science, which are the crème de la crème of the Indian nation.

The Indian Lunar Exploration program has included international cooperation from the beginning and will hopefully continue to do so in the upcoming Chandrayaan missions as well. Invitation to these types of prestigious cooperation programs could well serve as an appetizer for newcomers in the future. Additionally, India aims to demonstrate independent human spaceflight after 2020.

Although totally initiated and funded by the Indian Government to promote development, the program has many accomplishments, including the development of a home grown launch vehicle and indigenous satellite platform, boosting scientific interest, technological capability and public and institutional the awareness about the Moon within India. Moreover, providing a free-ride for international contributors has marked the Chandrayaan-1 initiative itself as one of the best and most successful opportunities to discover outer space together with other nations. It is the most international lunar spacecraft ever designed.

3.3 Others

Futron Corporation released the 2010 Space Competitiveness Index in which countries are ranked according to their space competitiveness, which was measured using a method developed by the company. The top 10 Countries (Europe being considered as one entity) and their ranks are reported as follows (Futron, 2011):

The first six countries have already been discussed up to this point. Three other countries that can be considered as newcomers are shortly introduced in following sub-sections.

Country	Rank
USA	1
Europe	2
Russia	3
Japan	4
China	5
India	6
Israel	7
South Korea	8
Canada	9
Brazil	10

Table 1. Top 10 in Futron's Space Technology Capacity Index.

3.3.1 Israel

Despite being the geographically smallest country among other newcomers, Israel reached its indigenous launch capabilities much earlier than many of the countries mentioned in this chapter. This success is based on its ballistic missile program in 1980's, and the help of a very strong local defence industry. Although recently many scientific applications have been developed, mainly by scientists originating from the Ukraine and Russian Federation, the main scope of the Israeli space program is defence needs and the country has no restrictions to use export licensed space products.

Due to geographic constraints, Israel is planning to launch its rockets from aircraft, similar to Indonesia and thereby avoid drop zone problems. Israel also cooperates with ESA via EU 7th Framework Programme (FP7) programs, Ukraine, the Russian Federation and also generally with the US.

3.3.2 South Korea

The South Korean Aerospace Research Institute, KARI, was founded in 1989 and so far has invested in earth observation, meteorology, communication and ocean monitoring satellites, launch vehicles and human space flight. Today a lunar lander prototype is ready and KARI would like to realize its lunar exploration program until 2025, following the successful qualification of the KSLV (Korean Space Launch Vehicle) rocket, many other spacecraft technologies, and procurement of necessary funds.

3.3.3 Brazil

The Brazilian space program, initiated in 1961, is primarily launch vehicle oriented. After several sounding rocket trials, a collaboration agreement with China was signed in 1988 resulting in the China-Brazil Earth Resources Satellite program (CBERS). So far, three satellites have been launched and two more are on the way. Brazil has also signed cooperation agreements with Canada, ESA, NASA, Russia, Ukraine and France and is also looking for partnership opportunities with Israel. The country has owned the Alcântara Launch Centre since 1982, and has collaboration programs with Ukraine based on the Cyclone-4 launch vehicle.

South Korea and Brazil clearly show promise as future players in space exploration, thanks to the political support from their governments, financial capabilities of their economies, and promising launch vehicles for independent access to Earth's orbit.

***

However, countries aspiring for space are not limited to the list given above. Many countries are already operating satellites, as shown in Figure 3. Some countries who are not contented with being the final users and operators of space systems created by a few industrialized countries, and who have a certain economic, demographic and technological capacity, have already initiated space programs to create their own space industry. The problem of establishing basic space technology capabilities with limited budgets and creating a sustainable, sound industry that can at least fulfil domestic needs is already well addressed in the literature (Leloglu & Kocaoglan, 2008; Jason et al., 2010; Waswa and Juma, 2012), so we do not discuss this topic in this study. The activities of some newcomers, namely South Africa, Thailand, Malaysia and Indonesia, which are in the category of newly industrialized countries, and some other countries are summarized as follows.

3.3.4 Taiwan

Taiwan is one of the more interesting examples with its Formosat satellite program and desire to develop its own indigenous launch vehicle. Unlike mainland China, they have no problem in procuring western products. The main aim of Taiwanese National Space Organization is to establish national self-reliant satellite technology. Taiwan, being technologically and financially more advanced than most of the newcomers with an export oriented economy, aims to develop local space technology infrastructure as well as to employ competitive resources that would favour Taiwan's space application industries for future international space markets. This would in turn benefit the development of space technology for basic daily needs, increase the breadth of scientific applications, and keep valuable human resources inside Taiwan, thereby increasing competitiveness and added value for domestic high technology industries such as telecommunications, nanotechnology, electronics and defence.

Due to the political balance in South East Asia, Taiwan generally allies with the United States and Europe, rather than pursue regional cooperation, and thus faces no obstacles inhibiting it from benefiting from International Traffic in Arms Regulations (ITAR) restricted US space technologies and launching its satellites via US military launch vehicles like Minotaur, Athena and Taurus. In this way, Taiwan has solved generic "procurement of export licensed qualified components" and "arrangement of launch campaign" problems.

3.3.5 South Africa

South Africa launched its first indigenous satellite, Sumbandilasat, in 2009, which continues to operate successfully. Future plans include establishing a space agency, and investing in launch vehicles and earth observation satellites. Although external funding for future projects is uncertain, South Africa aims to pursue its space-based goals with maximum local contribution and governmental support.

3.3.6 Turkey

Turkey started benefitting from space technologies in early 1990's by communication satellites. Towards the end of 1990, the country signed know-how and technology transfer agreement with the same source of Algeria and Nigeria. As a result, the first earth observation satellite, BiLSAT is launched in 2003.

So far, Turkey procured six more communication satellites, one earth observation satellite and also manufactured two more earth observation satellites at its own premises in the capital city, Ankara. One of the indigenously developed satellites, RASAT, was launched in August 2011. Among several space industry companies, TUBITAK UZAY, as an ambitious newcomer, is planning space exploration projects in the medium term in parallel to the National Space Research Program adopted by Supreme Council for Science and Technology in 2005 (TÜBİTAK, 2005).

In the meantime, some necessary capabilities are acquired; an on-board computer that can be used in interplanetary missions was developed. Some other hardware that can survive high radiation environment, including communication and power modules and vital software packages are in the process of development and project for establishing infrastructure and development environment for electric propulsion is being conducted. Hall Effect thrusters are being developed in parallel. An international project was also launched with Middle East Technical University and Ukrainian institutions.

TUBITAK UZAY also takes part in several EU FP7 projects and has submitted various projects with foreign partners. The European Cooperation for Space Standardization (ECSS) and Consultative Committee for Space Data Systems (CCSDS) standards and various established industry practices are followed to facilitate future international cooperation.

3.3.7 Thailand

Apart from the experiences of Thaicom in telecommunication satellites, Thailand's Remote Sensing Centre (GISTDA) ordered the first earth observation satellite, Theos, from France in 2004. Today, Theos still operates successfully to serve the daily imagery needs. In additional to five telecommunication satellites and one earth observation satellite, Thai universities have invested in Ka-band transponder development studies and several balloon experiments to observe the ionosphere both indigenously and in cooperation with the other South Asian nations participating in the Asia Pacific Space Cooperation Organization (APSCO) and Asia Pacific Regional Space Agency Forum (APRSAF) initiatives.

3.3.8 Malaysia

Malaysia is an interesting case as it established its space agency relatively recently in 2002 and has invested in human spaceflight, purely for prestige and public awareness. The first Malaysian astronaut visited the ISS under the Angkasawan program in 2007

The satellite experience of Malaysia started with a small technology transfer project and continued with complicated and operational Razaksat satellite project, which is a technology transfer from South Korea. The satellite was launched in 2009. Similar to Thailand and many other countries, Malaysia has invested heavily in commercial telecommunication satellites. The local telecommunication satellite operator has procured three satellites, similar to

Thailand, to serve communication applications and benefit from the financial return. Malaysia is an active member of APRSAF and collaborates not only with other nations in this organization on space technologies, but also works together with the Russian Federation on suborbital launch vehicle technologies.

3.3.9 Indonesia

Indonesia established the Indonesian National Aeronautics and Space Institute (LAPAN) in 1964 and has invested mainly in launch vehicle technologies with the help of the Russian Space Agency, and in the Palapa and Telkom telecommunication satellites, similar to its South Asian neighbours. Indonesian satellite development activities were initiated with the help and on-the-job training from Berlin Technical University and resulted in the development of two LAPAN mini satellites. The country signed a collaboration agreement with Ukraine in 2008 to study launch vehicle technologies. Additionally, Russia and Indonesia signed a commercial agreement that resulted in the construction of air-launch infrastructure on one of the islands in the Indian Ocean, one commercial communication satellite, and manufacturing of one other by Russia. LAPAN is currently indigenously developing a launch vehicle that is capable of delivering 100+ kg satellites to low earth orbit and two mini remote sensing & Automatic Identification of Ships (AIS) satellites in the 70 kg class.

3.3.10 Algeria

Similar to other countries, Algeria embarked on its space adventure via a technology transfer program from the Western Europe. Successful satellite design, test and operation experience of Alsat-1 satellite resulted in the construction of a satellite assembly integration and test (AI&T) facility in Oran city and the advanced Alsat-2A satellite, designed by a European company. After these technology and know-how transfer projects, Algeria is now developing its third satellite Alsat-2B at its own AI&T facility with its own personnel, and will be the third country in Africa to reach and benefit from space technologies, following South Africa and Nigeria.

3.3.11 Nigeria

Nigerian Space Agency launched its second and third remote sensing satellites together in August 2011; Nigeria ordered two more telecommunication satellites for commercial use with turn-key contract.

3.3.12 Egypt

Egypt was fortunate to have benefitted from a technology transfer program from Ukraine before the recent political depression and operates the Egypt-1 earth observation satellite.

3.3.13 Pakistan

Pakistan started with a technology transfer program from the United Kingdom and is now working with China for both turnkey telecommunication and earth observation satellite programs. Additionally, the Pakistan Space Agency is developing its own systems for testing on board the Chinese-made Paksat-1R communication satellite launched on August 2011.

3.3.14 Iran

Iran is developing its space technology mostly with the local resources. The country's launch vehicle program currently employs technologies enabling orbital distances of about 260 km and capable of carrying payloads in the 30 kg class. Although announcements about human spaceflight may not be realized in the foreseeable future, it is clear that Iran achieves more because of the embargo by developing applications and technologies with its own sources, rather than relying on technology transfer programs as other nations have done.

3.3.15 Vietnam

Vietnam is about the join the "Others" soon, with its two remote sensing satellites from France and Belgium, and a second telecommunication satellite from the US, currently on assembly.

Clearly, these latecomers are highly motivated and possess modest funding schemes mainly for "space for improving daily life" applications. Participation of the pioneers' space race with their local contribution will be valuable nationally and also encourage the rest of the world to join in this prestigious but very expensive work.

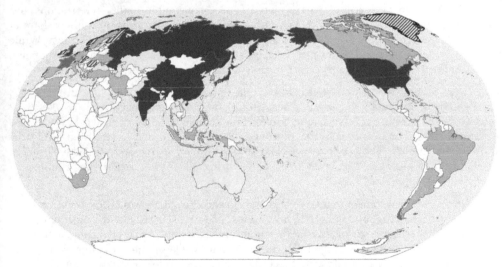

Fig. 3. Satellite operating nations. Dark blue: LEO, GEO and outer space, blue: LEO only (excluding cubesats), light blue: countries operating turnkey satellites systems

4. Newcomers in space exploration

In this section, we first visit the prerequisites of an ambitious program for space exploration. Then we discuss the major difficulties that an aspiring nation will face. Some advantages that the newcomers will enjoy are the subject of the following sub-section. Finally, we discuss some possible ways that these nations can contribute to space exploration with examples in the next section.

4.1 Prerequisites

Of course, certain prerequisites exist for the contribution of a nation to space exploration. The basic capabilities of space technology, including the infrastructure, like clean rooms and environmental testing chambers, human resources, basic know-how of design, assembly and test facilities, are musts. The newcomers that are the subject of this work are assumed to have already reached that level.

Another important prerequisite is the existence of strong universities and research institutes that can support scientific missions. This requirement is closely related to the science and technology policies and the R&D expenditures of the country, as well as strong GDP levels.

Per the Science Citation Index, many of the newly industrialized and developing countries are getting stronger and investing more to support their scientific and academic basis, which will be the main source of space science and technology studies.

According to a report published by the Royal Society in London (The Royal Society, 2011), China has acquired a second place ranking in the number of articles published in international science journals, and has already overtaken the UK. By 2020 China is positioned to take the leading position from the US. While the top 10 is still dominated by the major Western powers and Japan, who are producing high quality publications and attracting researchers to their world class universities and research institutes, their share of published research papers is falling and China, Brazil and India are coming up fast. While Western EU Countries and Japan produced 59 percent of all spending on science globally, their dominant position is nevertheless slipping against the newcomers.

The Royal Society report also states that China improved from sixth place in 1999-2003 (4.4 percent of the total) to second place behind the US over the years 2004-2008 (10.2 percent of the total), thus overtaking Japan. Newcomers like Iran and Turkey are also making dramatic progress. Turkey's improved scientific performance has been almost as dramatic as China's. The country increased its investment in research and development nearly six-fold between 1995 and 2007, and during the same period, the number of researchers increased by 43 percent.

The fact that the newcomers that have successful space programs are at the same time the countries whose share of scientific publications is increasing is not a coincidence. To summarize, achieving a strong scientific background relies on strong government funding for R&D and a sustainable budget for the universities. Newcomers lacking in a strong scientific and technological basis will have little chance to achieve success.

Key driving forces for the sustainability of space activities in the long term include, political will, public support, competitive pressure from neighbouring countries (Taiwan, Japan and Iran examples), in addition to basic capabilities and a strong scientific background. If we consider the emerging countries mentioned in the previous section, creation of public support shall be supported with providing employment opportunities for the new generation, supporting scientific opportunities for universities and institutes, answering daily   needs   like   disaster   management,   remote   sensing,   telecommunication, commercialization of developed technologies and funding spinoffs to step forward for industrialization.

Organizational capabilities are also important to succeed in space programs. An effective and efficient organization that can be kept out of daily political melee should coordinate all the efforts.

4.2 Difficulties

Although the world has seen an increase in space technologies and applications so far, many developing and newly industrialized countries have been facing several problems including "inadequate information, high cost, difficulty of accessing the data, no involvement of end users, sustainability of transferred technologies and lack of commercialization of space activities" (Noichim, 2008), limited availability of highly reliable, high performance, electrical, electronic and electro-mechanical components to trade restriction imposed by export licenses, agreements safe guarding technology, and US International Traffic in Arms Regulations (ITAR) and other countries' export licenses, technology safe guard agreements and dependency on other nations for launch campaigns. These generic and common problems are the basic hurdles in the race to space of newcomers. In this subsection, we summarize the most important obstacles to for space exploration missions for newcomers.

4.2.1 Access to space

Access to space is one of the major problems for newcomers to achieve orbital success. There is no doubt that certain countries may develop state of the art tools, payloads and spacecrafts, but only limited number of them are able to reach the orbit with their own will and abilities. Russia, the US and France are the main actors in this field and they inspired Asian nations, starting from India, China, Japan and South Korea in the area of space exploration. The first three have already reached sustainable, self reliant and self sufficient launch vehicle development programs to guarantee the access to space. However, the highly elliptical orbits necessary for outer space explorations may require more than the available capabilities of low earth orbit launch. Figure 4 summarizes the countries with the orbital launch capabilities, whose launch vehicles confirmed to reach orbit.

The experience of South Korea is a good example of difficulties of obtaining this capability. South Korea's Korean Space Launch Vehicle (KSLV) program was initiated with the cooperation of the Russian Federation and South Korea in 2004 as a part of a turn-key contract for the delivery of first stage engine of a launch vehicle, launch site and necessary services. South Korea contributed to the program with the second stage of the launch vehicle and test satellites. The KSLV is the first carrier rocket that made its maiden flight from Naro Space Centre in South Korea in 2009, followed by a second flight in 2010. Both flights dramatically ended up in failure and resulted in the loss of two technology demonstration satellites, moral, public support, motivation and public financing.

Today, Asian nations like Indonesia, Taiwan, Iran and Malaysia have sounding rockets or low earth orbit launch vehicle programs. Although Indonesia works with Ukraine and Russia while Malaysia works with Japan and Russia, there is a long way to go before these rockets serve space exploration missions.

Hence, most of the newcomers are dependent on launch vehicles from other countries. Dedicated launches for these missions are very costly and shared launches for the required peculiar orbits are very difficult to arrange and manage.

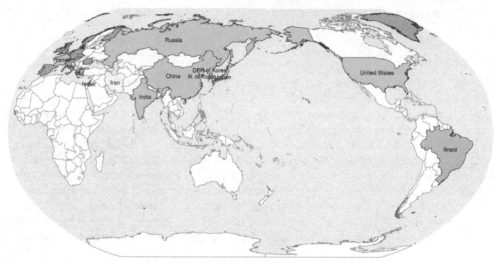

Fig. 4. Space Launch Capability. Pink: countries capable of launch technologies, Dark red: ESA, light blue: Countries with limited launch capability, blue: Countries thought to be very close to performing the first successful launch.

4.2.1.1 Important orbits for space exploration missions

The elliptical orbit is the primary way to access the Moon, Mars and beyond as they can provide escape from earth's gravity field. Orbits have different classifications from geostationary earth orbit (GEO) to geostationary transfer orbit (GTO), from Medium Earth Orbit (MEO) to Moon Transfer Orbit (MTO) or Earth-Moon Transfer Orbit (EMTO).

Low Earth orbit (LEO) is geocentric orbits ranging in altitude from 0–2,000 km and is the suitable orbit altitude for remote sensing satellites, suborbital launches, mobile communications, zero-g and biological experiments. LEO access is relatively more common than the others and number of LEO launch rockets and the number of countries who could achieve LEO access is more common.

Geostationary orbit (GEO) is the orbit around Earth matching Earth's sidereal rotation period. All geostationary orbits have a semi-major axis of 42,164 km. And this orbit is suitable for geostationary communications for TV, radio, telephone signals and meteorology applications while geostationary transfer orbit (GTO) is used for transferring communication satellites from LEO to GEO. GTO is an elliptic orbit where the perigee is at the altitude of a Low Earth Orbit (LEO) and the apogee at the altitude of a geostationary orbit. GEO launch vehicle is relatively limited than LEO launch vehicle and countries who could achieve this success are; US, Russia, China, France and India.

There are other certain orbit types that are used for outer space exploration. High Earth orbit (HEO) is the geocentric orbit above the altitude of geosynchronous orbit 3,786 km and suitable for escape manoeuvre via apogee kick engine and provides launch to solar system destinations except the Moon. They are also used for satellite radio applications by the US and communication purposes by Russia during cold war-era. HEO access is harder to achieve due to several cutting edge technologies onboard the launch vehicle and so far only Russia, US and Japan managed to launch satellites to HEO orbits.

To be able to reach the Moon, Moon Transfer Orbit (MTO-Hohmann transfer orbit) is generally used. In orbital mechanics, the Hohmann transfer orbit is an elliptical orbit used to transfer between two (typically coplanar) circular orbits. The orbital maneuver to perform the Hohmann transfer uses two engine impulses which, under standard assumptions, move a spacecraft in and out of the transfer orbit. This maneuver was named after Walter Hohmann, the German scientist. MTO orbit is achieved by Indian, Russian, Japanese, Chinese and US launch vehicles so far.

4.2.1.2 On board propulsion for space exploration

On-board propulsion is required to make necessary manoeuvres from the initial orbits for space exploration missions. Interplanetary travel requires new propulsion systems and new ways of generating power (Czysz, 2006). Although nuclear energy could be an alternative and unique way to discover our solar system and beyond, only Russia and the US have achieved this technology so far. This is a definitely limiting factor for newcomers wishing to pursue exploration of Mars and beyond. To be able to design satellites reasonably small to fit in launch vehicles, I_{sp}, the specific impulse, must at least double. However, a limiting factor for using chemical sources starts at this point and they do not permit benefitting from commonly used cold gas propulsion or hydrazine systems to be employed on board, as these chemical sources will finish much earlier than providing necessary thrust.

For outer space transportation, the ultimate alternative could be ion propulsion or Hall Effect thrust, which is a mature and qualified technology. This technology is safe, peaceful and easily be accessible for at least some of the newcomers and attracts the way for outer space exploration.

However, another major problem is maintaining the temperature of the satellite battery and other subsystems as the spacecraft grows increasingly distant from the Sun and the heating effect of the sunlight to approach other space objects like Mars. It is clear that a simple way will have to be discovered by scientists to solve this problem so that reliance on nuclear reactors for propulsion is terminated. Otherwise, all nations will remain dependent on the nuclear superpowers, which is another limiting factor for newcomers to pursue outer space exploration.

4.2.2 Funding

The next obstacle is the difficulty to convince politicians to allocate sufficient funds for space exploration. The funding of costly projects like telecommunication satellites, high-resolution earth observation satellites, or launchers is easier to justify on economic, strategic or security-related grounds. Although space exploration projects can be defended for their technological returns in the long run, spill-over effects, reversing brain-drain and promoting science, and their positive psychological effects on the public, securing the necessary funds is not easy. Most space exploration missions are extremely costly, for example, NASA's recent mission to Jupiter costed 1.1 billion US dollars. Most newcomers have difficulty in fronting that kind of expenditure. Even the Indian Space Agency, ISRO, who has had tremendous success in their space programs, is having difficulty to defend budget allocation for future Chandrayaan programs.

4.2.3 Dependency

Another basic problem for newcomers seems to be the dependency on other nations for specialized spacecraft technologies, such as radiation tolerance, propulsion technologies for complicated orbital manoeuvres, geographic distribution of ground stations networks, launchers, and the employment of international standards that are different than national ones. Unfortunately, many of these technologies are protected by national or well known international safeguards. Once a qualified space technology is protected and distribution is limited, newcomers are compelled to depend on other components, which may be less reliable or result in reduced performance, thereby slowing progress and increasing the risk in newly designed spacecraft.

4.2.4 Advantages

Although the space industry cannot be considered to be labour-intensive, the cost of recruiting the necessary high-skilled staff is an important component of space program costs. In developing and newly-industrialized countries, the labour costs of the engineers, scientists and other technical people are considerably lower compared to equivalent workers in developed countries.

In (Leloglu, 2009), the advantages of latecomers in space technologies have been discussed in detail. To summarize, some of the advantages are the ability to exploit literature published based on the difficultly-acquired experience of others; the accessibility of space equipment from various suppliers, which facilitates integration of space systems; a rich spectrum of technology transfer options; and developments in nano- and micro-satellites that enable the acquisition of basic capabilities with relatively modest resources.

5. Opportunities for space exploration

The most important mechanism for overcoming difficulties encountered along the way to realizing space exploration missions is international cooperation. Countries may share the costs and risks of expensive and ambitious projects. Partners may also benefit from complementary capabilities and geographic distribution of available ground stations. Another advantage pointed out by Petroni et al. (Petroni et al., 2010) is that collaboration enriches the capabilities of both sides by "exchange of knowledge and skills". Even the big space powers collaborate on several aspects of space explorations. For example, China and Russia worked together to explore Mars via the Phobos-Grunt program. While Russian Phobos Grunt is supposed to go to Mars, it would also provide a launch and transportation opportunity for the Chinese Mars orbiter Yinghuo-1. However, the satellite failed to leave Earth's orbit after launch.

Another mechanism for cooperation in space is the joint collaboration between a newcomer and an experienced agency. Taiwanese Formosat satellite project is a good example of this kind of cooperation. Formosat-1 and 2 spacecrafts and their payloads were developed jointly by the Taiwanese National Space Organization (NSPO), US and European suppliers and launched by US launch vehicles in 1999 and 2004, respectively. Formosat-3, aka, COSMIC (Constellation Observing System for Meteorology, Ionosphere, and Climate) was launched in 2006 and consisted of six spacecrafts. Taiwanese and US agencies not only shared the cost but also shared the data gathered from these ionospheric research satellites. On this project, Taiwan has mainly focused on payload development and benefitted from reliable, qualified US launch vehicles and the widely distributed US ground station network. Although only one of the satellites remains active today, the technology that Taiwanese institutions developed and hands on experience for NSPO employees paved the way for the Formosat-5 program, involving joint Taiwan-Canada-Japan collaboration and also resulted in an efficient use of resources.

The cooperation between ESA and other countries is another example. ESA has relationships with non-European countries such as Argentina, Brazil, China, Japan, India, Canada, US and Russia. Argentina is different from the others with respect to its space capabilities, however, the country benefits from its geographical location and supports ESA's future deep space missions to Mars and beyond. In return, ESA provides joint training courses for Argentinean students in various areas.

The EU Framework Programs is another example. The 7th Framework Program (2007 - 2013) is open to non-EU countries such as Turkey, Israel, Switzerland, Norway, Iceland, Liechtenstein, Croatia, Macedonia, Serbia, Albania, Montenegro, Bosnia & Herzegovina and the Faroe Islands. The FP7 Space Work Programme covers areas like "Space Exploration" and "RTD for Strengthening Space Foundations" (European Union, 2006). If these countries could succeed in becoming partners to space projects, in theory they would also be able to jointly develop key technologies. However, in practicality, it is not easy to take part in such projects due to the requirements of space heritage for products and compatibility with mainly ESA driven international standards such as The European Cooperation for Space Standardization (ECSS) and Consultative Committee for Space Data Systems (CCSDS).

Regional cooperation is another type of cooperation for which ESA is a very bright example. Two such initiatives in Asia are the APRSAF led by Japan and APSCO led by China. In these cases, at least one nation possesses launch vehicle capability and existing distributed ground station networks are needed.

Another example is International Space Exploration Coordination Group (ISECG) formed by 14 space agencies, namely Italian, French, Chinese, Canadian, Australian, US, UK, German, European, Indian, Japanese, Korean, Ukrainian and Russian space agencies in 2007. ISECG aims to formalize the vision for future robotic and human space exploration to solar system destinations, starting from the Moon and Mars, based on voluntary work approach and exchange information regarding the named space agencies' interests, plans and activities in space exploration with their "The Global Exploration Strategy: The Framework for Coordination" approach. ISECG is a good model for newcomers to pursue the way ahead for joint outer space exploration and be part of the coordination, basically to eliminate the duplication in this area.

On the other hand, regardless of the composition or existence of partners, there are technological solutions that can reduce costs or increase launch options. An important revolutionary mission is SMART-1, an ESA-funded satellite developed by the Swedish

Space Corporation. Using the French-made Hall effect thruster, the satellite could reach lunar orbit in more than one year from its initial geostationary transfer orbit. The Hall Effect thruster is in fact relatively old technology and has been in use since the 1960's in Russia. Although this technology is generally used in geostationary telecommunication satellites for station keeping manoeuvres, Smart-1 is one of the first examples of using the Hall Effect thrusters out of a geostationary earth orbit. Smart-1 has about 80 kg on board xenon and has managed to reach a total of 3.9 km/s ΔV in 5000 hours of operation. The spacecraft has demonstrated a cheaper, safer (with respect to hydrazine propulsion) version of space exploration by means of non-conventional propulsion technologies. Some standards designed for deep space communication that enable the reliable transfer for large amounts of satellite data over a very limited-bandwidth communication link by CCSDS, an international organization were also successfully qualified by Smart-1 and enabled future deep space missions to transmit larger volume of data back to earth from a distance of thousands and millions of kilometres away. In the final analysis, this mission provided very valuable experience to ESA and paved the way for the future, long, relatively cheap and safer missions to the Moon, Mars and beyond. The equipment qualified on board Smart-1, such as infrared and X-ray instruments, were also used in Indian lunar mission, Chandrayaan-1. Also, this mission enabled ESA to sign cooperation agreements with China, India, Japan, Russia and NASA regarding joint lunar programs.

Another groundbreaking and extraordinary example of a relatively low-cost space exploration mission to Mars was Beagle-2. The Mars Express Orbiter carried Beagle-2 to the orbit of Mars. Although the mission failed, it had the possibility of success due to strong support from ESA by means of ground stations, NASA by allowing a co-passenger on the mothership Mars Express, and Russian Space Agency with launch service support. Again, international collaboration was the only feasible way for this kind of space exploration mission. This was facilitated by a consortium set up by the project management office, and included universities and industry. After the development phase started, a European defence and space conglomerate took over the responsibility for managing the entire program. Thereafter, one of the most outstanding financial support campaigns was organized in which British pop music artists and painters were called upon to increase the awareness of the project in the public, mainstream media, and schools. In fact, the beacon signal of the spacecraft was composed by a British pop music band and several subsystems, including the cameras, were polished by a British painter to attract the attention of mainstream media. Given the enormous public support, the main ground control station was kept open to public to show where the funds had been used. Although the mission failed at the end, the Beagle-2 was used in several science fiction movies to strengthen the image that the spacecraft actually reached the planet Mars. Nevertheless, the Beagle-2 project continues to serve as a valuable example for how support from popular artists can be used to increase public awareness. For the first time, financial donations from ordinary citizens of all ages, wealth, and occupation were used to fund a space project, and as such Beagle-2 will always remain a unique project development success story.

In keeping with the low cost theme of the mission, the control software was the first of its type deployed on a laptop and several on board systems, which were not designed and manufactured with space qualification criteria, procured from the industry; similarly, mass spectrometer was provided by University of Leicester and University of Aberdeen.

These examples show that innovative solutions can be possible for the purpose of space exploration missions with limited resources. Newcomers can find novel creative solutions to realize their missions by optimizing their capabilities and cooperation opportunities. Moreover, cube-satellites and small satellites provide low-cost experimenting opportunities for scientific instruments, solar sails, formation flight technologies, tether tests and similar technologies.

As indicated by Petroni et al., to create an innovative mission to decrease the costs or increase the reliability, one important path that needs attention is to transfer technologies from non-space sectors and from the universities. (Petroni et al., 2010)

Another crucial way to communalize outer space exploration is to benefit from distributed, common ground stations and communication systems that are designed according to CCSDS protocols and standards to collectivize different systems can work in harmony and communicate with each other, especially on deep space missions, where spacecraft is seen commonly on the other side of the earth and throughout the day, forcing owners to use deep space ground stations owned by other countries.

For example, integration of Chinese, Indian, Russian, European and US Deep Space Networks via CCSDS standards could facilitate the achievement of distributed and sustainable outer space exploration, benefitting all mankind and eliminating duplication of individual efforts and unnecessary spending.

6. Conclusion

Space exploration has been a privilege for a few developed countries during most of the space age; however, as more nations get involved, space is becoming increasingly democratized. This has been made possible by technological developments as well as political changes as the global level. As the space programmes of nations new to space race advance, investments in space science and space exploration have increased, and, as a result, even more countries are getting involved. Although these new nations can benefit from the latecomer's advantages, they still need to overcome many obstacles to be able to contribute meaningfully to space exploration. There is a strong relationship between national science and technology policies, and advancement in space science and technology. Hence, investment in R&D backed by sound policies is a must for a successful program. Newcomers also need to seek international cooperation with strong space agencies and/or peers to share risks, costs and create synergy. Rather than imitating the missions of pioneers, they may try to find novel innovative solutions enabled by new technologies and an increasing number of international players and missions. Finally, aspiring nations should prepare for the future by following a sound but flexible plan.

7. References

Anokhin, A. V. (1924). *Muterialy po shamanstvy u altaitsev*, Akademija Nauk SSSR, Leningrad
European Union. (2000) Decision No 1982/2006/EC, *Official Journal of the European Union*, 30.12.2000
Futron Corporation. (2011). *Futron's 2010 Space Competitiveness Index*, www.futron.com (access: 10 April 2011)

Gömeç, S. (1998). Shamanism and Old Turkish Religion (In Turkish), *Pamukkale University Journal of Education*, No.4, pp. 38-52, ISSN 1301-0085

Jason, S., da Silva Curiel, A., Liddle, D., Chizea, F., Leloğlu, U. M., Helvacı, M., Bekhti, M., Benachir, D., Boland, L., Gomes, L. & Sweeting, M. (2010). Capacity Building in Emerging Space Nations: Experiences, Challenges and Benefits. *Advances in Space Research*, Vol.46, No.5, (September 2010) pp. 571-581, ISSN 0273-1177

Leloglu, U. M. (2008). The Small Satellite Odyssey of Turkey, *International Workshop on Small Satellites, New Missions and New Technologies, SSW08*, Istanbul, June 2008

Leloglu, U. M. (2009). Latecomer Advantage In Space Technologies: A Posse Ad Esse, *Data Systems in Aerospace, DASIA 2009*, Istanbul, Turkey, 26-29 May 2009, CD-ROM

Leloglu, U. M. & Kocaoglan, E. (2008). Establishing Space Industry in Developing Countries: Opportunities and Difficulties. *Advances in Space Research*, Vol.42, No.11, (December 2008), pp. 1879-1886, ISSN 0273-1177

Logsdon, J. M., (2008), Why space exploration should be a global project. *Advances in Space Research*, Vol.24, No.1, pp. 3-5, ISSN 0273-1177

Noichim C., (2008). Promoting ASEAN space cooperation. *Space Policy*, Vol.24, No.1, (February 2008), pp. 10–12, ISSN 0265-9646

Petroni G., Venturini, K. & Santini, S., (2010). Space technology transfer policies: Learning from scientific satellite case studies. *Space Policy*, Vol.26, No.1 (February 2010) pp. 39-52, ISSN 0265-9646

The Royal Society, (2011) *Report; Knowledge, Networks and Nations: Global scientific collaboration in the 21st century*, London, March 2011

TÜBİTAK, 11th Meeting of Supreme Council for Science and Technology Decision Report. (2005). 63-100, 10 March 2005. http://www.tubitak.gov.tr/tubitak_content_files/BTYPD/btyk/11/11btyk_karar.pdf

Waswa, P. & Juma, C., (2012). Establishing a Space Sector for Sustainable Development in Kenya, *International Journal of Technology and Globalisation*, Vol.6, No.1/2, ISSN 1476-5667

Czysz, P. A. & Bruno, C., (2006) *Future Spacecraft Propulsion Systems: Enabling Technologies for Space Exploration*, Springer-Praxis, ISBN 10: 3-540-23161-7

Woellert, K., Ehrenfreund, P., Ricco, A.J., Hertzfeld, H. (2010). Cubesats: Cost-effective science and technology platforms for emerging and developing nations. *Advances in space Research*, Vol 47, No.4 (October 2010), pp. 663-684, ISSN 0273-1177

Part 2

Evolution of the Earth

Why Isn't the Earth Completely Covered in Water?

Joseph A. Nuth III[1], Frans J. M. Rietmeijer[2] and Cassandra L. Marnocha[1,3]
[1]*Astrochemistry Laboratory, Code 691 NASA's Goddard Space Flight Center, Greenbelt*
[2]*Dept. of Earth and Planetary Sciences, MSC03-2040*
University of New Mexico, Albuquerque
[3]*University of Wisconsin at Green Bay, Green Bay*
USA

1. Introduction

There is considerable discussion about the origin of Earth's water and the possibility that much of it may have been delivered by comets either within the first several hundred million years or possibly over geologic time [Drake, 2005]. Typical models for the origin of the Earth begin by assuming the accumulation of some combination of chondritic meteorites (Javoy, 1995; Ringwood, 1979; Wanke, 1981), yet it is highly likely that the asteroids that went into the terrestrial planets are no longer represented to any significant extent in the present-day asteroid population (Nuth, 2008). The argument concerning the composition of the building blocks of the Earth is typically phrased in terms of the chemical composition of the terrestrial mantle as compared to that of primitive meteorites [Righter et al., 2006] or to the isotopic composition of the Earth's oceans as compared to that of cometary water [Righter, 2007]. Both of these considerations yield important constraints on the problem. However, we will demonstrate that a completely novel examination of the problem based on models of nebular accretion, terrestrial planet formation and the evolution of primitive bodies makes using any modern meteorite type as the basis for understanding the volatile content of the Earth inappropriate. Unfortunately, while this approach yields terrestrial planets with sufficient water to easily explain the Earth's oceans, it also introduces a new problem: How do we get rid of the massive excess of water that this model predicts?

The mechanism for the formation of the terrestrial planets has been the subject of considerable debate. Gravitational instabilities in a dusty disk (Goldreich & Ward, 1973; Youdin and Shu, 2002) may have been responsible for planetesimal formation on very rapid timescales compared to the lifetime of the nebula. On the other hand, collisional accretion of larger aggregates starting from primitive interstellar dust grains should also occur in the nebula (Blum, 1990; Blum and Wurm, 2000), and these aggregates could continue to evolve into kilometer scale planetesimals or even into proto-planetary scale objects. While it is not clear if dust and gas can be concentrated sufficiently to trigger the gravitational accretion (Cuzzi and Weidenschilling, 2006) of planetesimals or proto-planets, it is clear that some level of collisional accretion must occur in proto-planetary nebulae in order to at least make chondrule precursors and probably to make components of meteorite parent bodies, meters to tens of meters in diameter. Youdin and Goodman (2005) have suggested that turbulent shear instabilities might serve to concentrate mixtures of chondrules and dust to sufficient density for gravitational instabilities to directly form meteorite parent bodies. As these bodies grow from dust grains to larger sizes, they also drift inward due to gas drag and can be lost into the sun in about a century (Youdin, pers. comm.). This drift might also serve to bring ices to the terrestrial planet region from beyond the snowline, and could thus be a primary source of water that has not been accounted for in models of planet growth. The purpose of this effort is to investigate the potential for such a mechanism to deliver water to the terrestrial planets as they grow.

Weidenschilling [1997: Hereafter W97] published an excellent model for the formation of comets in a minimum mass solar nebula. In this model, because the growing icy agglomerates slowly decouple from the gas as they gain mass and become more compact, comets begin to form at nebular radii between about 100 - 200 A.U. and fully decouple from the gas at 5 to 10 A.U. having grown into planetesimals on the order of 10 - 15 km in diameter. In more massive nebulae, the feeding zone for materials incorporated into a growing planetesimal would be proportionally smaller and icy agglomerates that begin accreting at 200 A.U. might easily reach diameters of 10 - 15 km before leaving the region of the Kuiper Belt. We have tried to extend a very simplified version of this general model to delineate the feeding zones for planetesimals that might have been accreted into the early Earth.

In the sections that follow we will describe our very simple calculations of the feeding zones from which the planetesimals formed that aggregated into protoplanets at 1 A.U. We will discuss the results of these calculations in terms of their dependence on the mass of the solar nebula and demonstrate that if this accretion model is applicable to even some planetesimals, then the Earth may have accreted an enormous quantity of water, the majority of which must have escaped early in our planet's history. We will very briefly discuss reasons why meteorites in our modern collections are unlikely to represent the materials that accreted to form our planet 4.5 billion years ago, and how our modern sample is likely to differ from these more primitive materials. We will also discuss possible alternative scenarios for the accretion of the proto-Earth, and how these scenarios might change our conclusions.

2. Methods

Weidenschilling (W97) published what we consider to be one of the best models for the formation of comets in a low-mass nebula. The model is based on the simple accretion of nebular solids and the effect that increased mass has on an object in orbit about the sun in a gas filled nebula. Specifically, tiny solids are initially very closely coupled to the gas. As accretion proceeds, a growing body begins to orbit independently and decouples from the gas. However, while the local gas and dust is partially supported by gas pressure, requiring a lower velocity to maintain its position in the nebula, a growing dusty snowball soon begins to drift inward without the support of the surrounding gas. This drift brings the accreting planetesimals into contact with fresh materials across a broad feeding zone, and continues until the body has grown to a size where its orbital velocity is no longer significantly affected by gas drag. Weidenschilling's (W97) purpose was to identify the radial dimensions of the nebula that produced the comets that were scattered by the giant planets into the Oort Cloud. It is our intention in this work to extend Weidenschilling's results to determine the composition of the population of planetesimals at 1 A.U. from which the terrestrial planets may have formed.

Using Weidenschilling's results together with Hayashi's model for the solar nebula (Hayashi, 1981), based on the minimum mass solar nebula, we were able to construct a simple model maintaining the nebular structure used in both authors' works. Hayashi's model provides equations for the surface density of gas, rock, and solids as a function of annular distance. The surface density for rock (ρ_r) is given as $7.1\ r^{-1.5}$ g cm^{-2} and the surface density for solids (ρ_s) as $30\ r^{-1.5}$ g cm^{-2}. These equations also apply to the assumptions of surface density at 30 AU used in Weidenschilling's model. Assuming that the composition of the nebula does not change as we increase nebular mass, then the "surface density" phase coefficient for rock or solids in higher mass nebulae is easily calculated by increasing each by the same factor by which one wishes to increase the nebular mass. These surface density coefficients are used to calculate the mass of either rock or solids (ice plus rock) between two given points via the following equation:

$$\text{Total Mass (g)} = 4\ \pi\ Q\ [r_b^{1/2} - r_a^{1/2}][1.496 \times 10^{13}]^2 \tag{1}$$

where Q is the phase coefficient for a given nebular mass in units of g cm^{-2}, $[1.496 \times 10^{13}]^2$ converts(A.U.)^2 to cm^2, r_b is the annular distance in A.U. at which the planetesimal begins accreting material, and r_a is the annular distance at which the planetesimal ends its accretion (presumably merging into a larger body).

Weidenschilling (W97) used numerical methods to calculate that a planetesimal must start at about 200 AU in order to grow to 15 kilometers by the time it drifts in to 5 - 10 A.U. Using the solid surface density coefficient ($\rho_s = 84$) for Weidenschilling's model, as well as his stated "start" and "end" points, we calculate that a planetesimal must travel through 2.494×10^{30} grams of material in order to accrete to 15 km. This equates to an accretion efficiency of ~ 1.41 parts in 10^{12} depending on the density assumed for the accreting planetesimal. This "accretion rate" is used throughout our calculations such that: planetesimal mass (in g) = total mass of solids and/or rock between r_b and r_a /1.41 x 10^{12}, or

$$\text{Mass (g)} = (4\ \pi\ Q\ [r_b^{1/2} - r_a^{1/2}][1.496 \times 10^{13}]^2)/1.41 \times 10^{12} \tag{2}$$

Note, we have assumed a density of 1.0 g/cc for all planetesimals. Determining approximate percentages of rock and ice found in a planetesimal was dependent on the starting and ending radii in reference to the snowline (which in this very simple model we have assumed to remain at 5 A.U.). Planetesimals beginning and ending accretion inside the snowline will be completely dry and 100% rock. Planetesimals that begin to accrete beyond the snowline and ending their accretion within the snowline must have the total mass (e.g., rock) from the snowline to the end point calculated, as well as the mass of solids (e.g., rock and ice) calculated from the starting point to the snowline. Again, using the surface density coefficients allows us to "separate" the mass of rock and ice accreted beyond the snowline from the pure rock accreted within, and thereby calculates approximate percentages for each.

Several important points must be made concerning our use of Weidenschilling's results before we begin to discuss our own. First, Weidenschilling reported extensive results from numerical simulations that explicitly took into account a wide range of factors such as the sticking of grains to a growing body. We have lumped all of these effects into an efficiency factor and have used this same efficiency factor to represent the accretion of both icy dust (outside the snowline) and dry rock (inside the snow line). This is obviously an oversimplification of a very complex and poorly understood process. Second, other processes may have created planetesimals in the solar nebula such as gravitational instabilities or large-scale vortices. Our study only examines the results of hierarchical accretion. Third, we assume that the snowline represents a discontinuity between a mixture of dry dust and hydrous gas inside the line and a mixture of dry gas and icy dust on the outside of this sharp divide. This obviously neglects the bodies that may have accreted to ten-meter or even kilometer scales outside this divide, yet drifted inward to some extent due to gas drag or gravitational interactions. Finally, we have assumed that the position of the snowline (at 5 A.U.) does not migrate as we increase the mass of the nebula, but instead remains fixed no matter how we change the mass of the system.

The effects of each of these simplifications will be examined below, after we have presented the results of our calculations. However, we contend that the effects of many of these factors would not tend to favor the accretion of rock over ice into planetesimals at 1 A.U. and that the uncertainties in nebular conditions, particularly in the mass of the nebula itself, are sufficiently large that our simplified calculations are an appropriate first step in examining the potential incorporation of ice and water into the progenitors of the terrestrial planets.

3. Results

From Weidenschilling's work [W97] we calculated an efficiency factor for the accretion of primitive planetesimals based on the final diameter of the body, the density of nebular solids (including ice) and the total mass of the nebular disk. We first validated this efficiency factor by using it to reproduce other examples of accretion calculations shown in Weidenschilling's (W97) paper. We then used this efficiency factor to calculate the size of the initial feeding zone for planetesimals that had grown to 10, 15 and 20 km in diameter by the time they reached 1 A.U. as a function of nebular mass. In other words, assuming that 10, 15 and 20 km sized planetesimals were present at 1 A.U., were still drifting inward but were available to be incorporated into growing protoplanets, where did these bodies begin to accrete? The results of our calculations are presented in Table 1 where we show the nebular radius where accretion begins as a function of the diameter of the planetesimal at 1 A.U. and the mass of the solar nebula. In all cases we assumed that the snowline is located at exactly 5 A.U. As can be seen from Table 1, the size of the planetesimal feeding zone and the percentage of ice in the final planetesimal are strong functions of the nebular mass and the size of the planetesimal itself: higher mass planets in lower mass nebulae contain much more ice.

We based our calculations on the total nebular mass expressed in units of the Hayashi Minimum Mass Nebula [Hayashi, 1981, Hayashi et al., 1985] and for each nebular mass we calculated where aggregation must begin in order to produce the planetesimal size of interest based on the accretion efficiency discussed above. We note that Weidenschilling [W97] used a value 2.8 times the Hayashi Minimum Mass in his model of comet formation, and more recently, Desch [2008] has estimated that the primitive solar nebula must have been at least 25 times the Hayashi Minimum Mass, but with a somewhat steeper slope (e.g. less mass in the outer regions of the nebula), thus more centrally concentrating the material available for planet formation. In this scenario, the snow line may be as close as 2.8 A.U. from the protosun. We decided to adopt the more conservative 5 A.U. position for the snow line in this work, and still only in the most massive nebulae do the smallest planetesimals at 1 A.U. contain pure rock. All planetesimals 10 km in diameter and larger contain a significant fraction of ice.

	Total Nebular Mass (Hayashi Minimum Mass Nebula)					
	1	10	20	30	40	50
Planetesimal Diameter (km)	Distance from the Proto-Sun where Aggregation Begins (A.U.)					
10	122	8.1	5.8	5.05	4.71	4.52
15	1068	25.2	12.1	8.81	7.36	6.55
20	5594	85.2	31.2	19.11	14.17	11.56
	Percentage of Ice in the Final Planetesimal (%)					
10	74	52	27	3	0	0
15	76	69	62	55	47	40
20	76	73	70	67	64	61

Table 1. Radius of Feeding Zone and Percentage of Ice in the Final Planetesimal as a Function of Nebular Mass and Planetesimal Diameter

4. Effects of our assumptions on the results

As noted above we made a number of simplifying assumptions in our calculations that could affect our results. First, we assumed that the snowline remains at 5 A.U. no matter the mass of the nebula. However, since any likely accretion scenario for the Solar Nebula would have a mass considerably larger than the Hayashi Minimum Mass, the snowline for more massive nebulae would occur closer to the proto-sun. This would tend to increase the number of planetesimals containing ice and increase the ice content as a fraction of planetesimal mass for small bodies at 1 A.U. In fact Desch (2008) calculated that the snowline for a nebula 25 times the Hayashi Minimum Mass would occur at 2.5 A.U., well inside the outer asteroid belt. Under these conditions, some outer main belt asteroids would certainly contain significant quantities of ice in their interiors.

Second, we assumed that we could extend the results of Weidenschilling's (W97) numerical calculations from the Outer Planets region to the Terrestrial Planets region and that the efficiency for particulate aggregation would remain unchanged. We find that reasonable changes in the efficiency factor used in these calculations do not change our basic results that smaller planetesimals at 1 A.U. accreted in more massive nebulae contain more rock while larger bodies in lower mass nebulae contain much more ice. Certainly the detailed results are modified with changes to this factor; however, the uncertainty in the actual mass of the Solar Nebula is more important in determining the ice content of the planetesimals than is the exact value of the accretion efficiency factor employed in the calculations. We also use the same factor for the accretion of both icy and anhydrous dust. The relative sizes of the particles and the velocities of the collisions appear to be more important than the exact composition of the accreting material, but the modest experimental results that we have to date indicate that collisions between icy particles are more likely to result in sticking than are collisions between dry rocks and pebbles. Our calculations therefore may overestimate the efficiency of forming ice-free planetesimals.

Third, we assumed that the snowline represents a sharp compositional boundary in the nebula. Accretion entirely inside the snowline will never incorporate ice into the planetesimal under this assumption. However, in the current solar system there are numerous examples of short period comets making many passes inside the orbit of the Earth, and subsequent apparitions of these comets demonstrate that they continue to retain at least some water vapor over thousands of years. Comets are even known to make several passes through the solar corona without complete loss of the ice in their interiors. The early nebula between 1 and 10 A.U. was certainly more opaque to solar radiation than our modern solar system and was more densely populated by icy planetesimals. Our assumption of a sharp discontinuity in the availability of small icy bodies inside the snowline that might be incorporated into planetesimals, even ones formed by turbulent accretion or gravitational instabilities, certainly favors the formation of ice-free planetesimals.

On each occasion where we made an assumption to simplify our calculations while still using the basic results of Weidenschilling (W97), we tried to consistently err on the side of producing the largest fraction of ice-free planetesimals possible. In spite of our obvious bias in favor of such dry dusty bodies, we nearly always produced a population of ice-rich planetesimals at 1 A.U.

Alternative accretion models

Weidenschilling's (W97) manuscript describing his model for planetesimal (comet) aggregation acknowledges that as the total mass of the Solar Nebula increases, additional factors, such as gravitational instabilities and other collective effects could increase the efficiency of accretion. In another treatment of the formation of the terrestrial planets Kokubo and Ida (2000) find that the formation of proto-planets occurs as a two stage process. Runaway accretion first produces a mixed distribution of planetesimals and proto-planets (Kokubo and Ida, 1995). The proto-planets initially grow rapidly at the expense of the planetesimal population (Kokubo and Ida, 1996) until a small number of proto-planets have formed in the nebula. These proto-planets then grow much more slowly, entering the "oligarchic growth" stage (Kokubo and Ida, 1998) where the larger proto-planets actually grow more slowly than smaller ones. In this scenario, proto-planets with masses of 10^{26} g are formed within about 500,000 years at 1 A.U. (Kokubo and Ida, 2000) and thereafter maintain rough separations greater than 5 Hill radii from one another as they continue to accrete planetesimals from their surroundings. These proto-planets form rapidly from young planetesimals that have not had sufficient time to lose any ice that may have been sequestered within, and such proto-planets are sufficiently large that they should retain a reasonable fraction of their accreted volatile complement.

The oligarchic growth stage from proto-planet to planet is a result of the larger bodies accreting materials from their immediate vicinity. While casual readers of these papers might believe that because these proto-planets incorporate materials from only a very narrow range of heliocentric distance, this must imply that the proto-planet that eventually grew to become the Earth could only incorporate rocky material. This impression is incorrect for two reasons. First, the initial distribution of planetesimals from which the population of proto-planets evolved was shaped by gas drag as the planetesimals accreted from the disk. Thus the population of planetesimals from which a proto-planet at 1 A.U. accreted should have been similar to that shown in Table 1.

Second, although the models of Kokubo and Ida (2000) span only a very narrow ring within the nebula (Δa = 0.02 – 0.09 A.U.) the boundary conditions (see their section 3.2) that they apply to these calculations assume a free flow of planetesimals through the model annulus. In other words, as proto-planets grow from the population of planetesimals within an annulus, the model assumes a balance between the inward loss of planetesimals due to gas drag and those that flow into the accretion zone from the outer nebula such that the surface density of the accreting annular disk remains constant. Thus, while the growing proto-planets remain stationary in this scenario, the planetesimal population from which they accrete continuously flows in toward the sun due to gas drag. Again the ice to rock ratio shown in Table 1 would apply to this population.

Finally, there is the possibility that some planetesimals could form on very short timescales due to turbulent-gas-driven gravitational instabilities. In turbulent eddies the local conditions for gravitational instability can be met (Johansen et al., 2006; 2007). The numerical simulations by Johansen et al. show the formation of Ceres-mass planetesimals in a few orbital periods. Also, chondrule-size particles can be concentrated in the low-vorticity regions of the disk (Ormel et al., 2008) and this can allow the formation of 50-100 km planetesimals on a short timescale due to the self-gravity of the chondrule clump. In both cases, the short timescales will prevent any significant radial migration. These mechanisms could easily produce relatively dry proto-planets if all of the components within the gravitational instability had equilibrated within the snowline.

While such processes would aggregate all of the mass of an individual planetesimal or proto-planet within a short timescale and from a very narrow range of heliocentric distances, the composition of such bodies will depend on the composition of the material that one would expect to be present at 1 A.U. while the composition of the proto-Earth would depend on the fraction of planetesimals that form via gravitational instabilities. Gravitational instabilities are unlikely to be the dominant process forming comets in the outer nebula as both the degree of MHD driven turbulence and the surface density of the disk are too low for efficient operation of this accretion mechanism. This argues that collisional aggregation processes that formed ever larger solid bodies that drifted sunward due to gas drag probably dominated in this regime (W97). If such processes dominated in the outer solar system, there is no reason to believe that this mechanism did not also operate at least to some extent in the inner solar nebula. In contrast, there are meteorite samples from asteroids that contained very little water, even at the time of their formation. As almost any sized body ending accretion near 3 A.U. would contain some ice if formed via the collisional-aggregation, gas-drag scenario, this provides evidence for the role of turbulent accretion or gravitational instabilities in the production of some planetesimals.

We can make the simple approximation that planetesimals formed via gravitational instabilities at 1 A.U. always consist of rocky materials. If half of all planetesimals at 1 A.U. formed via this mechanism, then the ice fraction of materials accreting to form the proto-Earth shown in Table 1 would be reduced by a factor of 2. However, as gas drag mediated migration of meter-to-kilometer scale planetesimals will always occur to some degree, some fraction of the larger planetesimals formed via gravitational instabilities will contain ice that drifted inward within bodies that began to accrete beyond the snowline. While small bodies eventually lose their volatiles in the inner nebula, rapid gravitational accretion could trap considerable fractions of this ice in larger, more robust planetesimals before it was lost. Therefore unless all planetesimals at 1 A.U. formed via gravitational instability and unless there were no small bodies present at 1 A.U. that began accretion beyond the snowline to be incorporated into these planetesimals, then the proto-Earth would have accreted considerably more water than is assumed in models that begin with planetesimals of chondritic composition.

5. Implications of the results

Young planetesimals were wet

Because many of these bodies began accreting well outside the snowline [Lunine, 2006; Ciesla and Charnley, 2006], they contain considerable quantities of water as ice grains, much like comets. However, can this water be retained to be incorporated into the growing Earth? In a dense nebula, the light of the protosun is unlikely to drive the loss of volatiles from a small planetesimal within the snowline as our sun does today. Instead, volatile loss was more likely controlled by the internal heat generated within the planetesimals via radioactive decay or by accretional impact. For small planetesimals that coalesce from even smaller bodies in similar orbits, impact energy is likely to be localized and relatively insignificant on the global scale. Such collisions resulted in planetesimals that we call comets when they now arrive from the outer solar system, and are unlikely to result in extensive volatile loss in the inner solar system.

The concentration of radioactive elements initially available for incorporation into the terrestrial planets depends to a large degree on how they were added to the system. If injection of such material initiated the collapse of the nebula [Wadhwa et al., 2006], or if they were simply present in the collapsing molecular cloud that formed our solar system [Chabot and Haack, 2006], then all planetesimals would accrete from roughly the same mix of material. If the short-lived radioactive elements were injected into the nebula at some time after nebular collapse [Wadhwa et al, 2006], then later formed planetesimals could contain higher fractions of these heat sources than planetesimals formed from the less radioactive solids in the molecular cloud core, and these younger bodies would therefore evolve faster than those formed earlier. However, even small (5 km) planetesimals that are enriched in ^{60}Fe and ^{26}Al require from a few hundred thousand to several tens of millions of years to reach their maximum internal temperatures [Das and Srinivasan, 2007] and become totally dehydrated. If such bodies contained substantial quantities of water as ice, then heating would be slower due to the reduced concentration of radioactive heat sources, and the slow loss of volatiles would allow these planetesimals to sweat, thus efficiently losing heat from their interiors.

While planetesimals heat slowly [LaTourette and Wasserburg, 1998; Huss et al., 2006], protoplanets form rapidly in a runaway growth process caused by the increased gravitational cross sections of larger planetesimals [Wetherill and Stewart, 1989; 1993]. It has been suggested that terrestrial planets may have formed within 10 million years of nebular collapse [Jacobsen, 2003; Jacobsen et al., 2009] and that core formation on the Earth occurred less than 20 million years later [Nichols, 2006]. In other words, the Earth may have formed from very young planetesimals that had not yet had a chance to lose any significant quantity of ice or water of hydration due to radioactive decay driven heating. In this scenario, a proto-Earth may have formed directly from the planetesimals whose ice content is listed in Table 1. This would result in a planet with much too much water, rather than too little. Given other considerations, the rocky fraction of the planet would most likely need to be comparable to the present mass of the Moon or Mars, and so the proto-Earth may have been nearly twice as massive if much of the incoming ice remained trapped within the growing protoplanet, on its very wet surface, or within its massive, water - rich atmosphere.

Composition of planetesimals

The composition of the rocky component of the planetesimals formed via gas-drag mediated accretion should be roughly chondritic. There is no mechanism for metal-silicate fractionation to operate on the dust grains and boulders that would be accreted via this mechanism. Although the overall composition of the planetesimals formed in this scenario bear more resemblance to what we currently call comets than to modern asteroids, for reasonable values of the nebular mass, the majority of the water accreted into these planetesimals originates in the inner solar system within a few A.U. of the snowline. For this reason, the planetesimals would not be expected to have the high D/H ratios or any significant content of volatile organic materials found in modern Kuiper-belt or Oort cloud comets.

As both the refractory composition of the planetesimals and the isotopic composition of the water would be "normal" when compared to typical models for the formation of the Earth, the only real compositional difference between the scenario described above and current models for the origin of the terrestrial planets is in the total quantity of water that might have been accreted. While current models often require the delivery of water at the end of the accretion process, this scenario requires the loss of water from the proto-Earth to be compatible with the composition of the modern Earth.

We do not have samples of the population of planetesimals that accreted to form the terrestrial planets in our modern meteorite collection [Drake and Righter, 2002]. Such planetesimals would have lost their ice and much of their water of hydration several billion years ago [Nuth, 2008]. The residual dehydrated body that began with more ice than dust and thus contained a smaller radioactive heat source than that which produced the large scale melting and differentiation of Ceres, Vesta and other asteroids would also be much more fragile than such solid rocks, much like loosely compressed sandstone. Collisional processes over the lifetime of the solar system would have gradually reduced the surviving number of these fragile bodies to an insignificant fraction of the asteroid population. It is therefore very likely that we do not have any representative samples of the planetesimal population that contributed to the formation of the terrestrial planets in our modern meteorite collections.

Loss of volatiles from the Earth

We expect that large quantities of water may have been lost from the growing proto-Earth due to impact induced heating, especially considering the lower escape velocity of the less massive, but rapidly growing protoplanet. We must also assume that the rocky terrestrial interior would remain at least fully saturated by water dissolved within the rocks and magma. In fact, given the overburden and likely difficulty of escape to the surface, we would expect that water vapor would become supersaturated within the proto-planet, and that any terrestrial proto-planet in the late stages of accretion would have a thick, water-vapor-laden atmosphere that should undergo some loss of water back to space. In the case of the Earth it is likely that this entire atmosphere, as well as any nascent hydrosphere, was lost during the impact of the Mars-scale body that formed the Moon [Cameron and Benz, 1991; Canup and Asphaug, 2001; Canup, 2004]. In addition, such a large impact would dehydrate a significant fraction of the terrestrial mantle as well as virtually all of the material from the impactor that might fall back onto the surface of the Earth. The Late Heavy Bombardment would have further dehydrated the terrestrial crust and uppermost mantle due to impact induced heating. Meanwhile the planetesimals impacting the Earth at this late stage would have had 500 million years since their accretion to lose a large fraction of their own volatile content, probably resulting in a net loss of water from the Earth.

If runaway growth morphed into oligarchic growth and only formed proto-planets in size ranges somewhere between Ceres and the Moon, then it may be much easier to explain volatile loss during the growth of the Earth. Because the initial proto-planets would be small, they would more easily lose volatiles that reached the surface than would a 90% finished Earth. In addition, a large number of collisional aggregation events are required to increase the mass of the Earth to its present value (~6 x 10^{27}g) starting from a population of much smaller objects (~10^{26}g). Each of these events would strip away any nascent hydrosphere on the colliding proto-planets, and could partially dehydrate the interior of the resultant body. However, if the interior of the growing proto-Earth remained near saturation (~3% water by mass) throughout the accretion process, the resulting planet would have many times more water than is needed to form the current hydrosphere (<0.025% of the Earth by mass). In fact, the above statement would still be true even if 95% of the water contained in all of the saturated proto-planets that accreted to form the Earth were lost in the growth process.

Water on the newborn Earth

Comets still impact the Earth today (e.g., Tunguska) and the frequency of such impacts was higher in the first billion years of Solar System history [Chyba, et al., 1994]. Comets therefore must have contributed some fraction of the water currently on the modern Earth. But we submit that a significant hydrosphere and a very wet atmosphere already existed on the proto-Earth prior to the Moon forming event, due to the accretion of ice laden planetesimals. Since up to 3% water can be dissolved in the modern terrestrial mantle at equilibrium [Righter, 2007] and even more water could have been present "in transit" through the mantle and crust as water worked its way up to the surface from the deeper planetary interior, we see no possible alternative but to accept that the early Earth had quite a large complement of water and may have been a bit more massive than expected when the moon forming collision occurred.

One might ask what the modern Earth would have been like if the Moon-forming event and the Late Heavy Bombardment had never occurred. The Earth's surface is already 75% ocean even though water comprises much less than 0.1% of the Earth's total mass. Had the early Earth not lost its original wet atmosphere, hydrosphere and some very large fraction of the water dissolved in its upper mantle, the entire surface of the Earth might today be covered by water to depths of at least several hundred miles, assuming that natural atmospheric erosion would have eliminated a substantial fraction of the initially accreted ice. Even with the moon forming event and Late Heavy Bombardment, the interior of the planet should still be rich in dissolved water and hydrogen: dissolved hydrogen could certainly be plentiful in the Earth's core, and most mantle magmas should be fully saturated in water.

Water equilibrates oxygen isotopes in the Earth-Moon system

If the proto-Earth and the "Mars-size impactor" were both equilibrated bodies containing significant reservoirs of liquid water prior to impact, the energy of the collision would vaporize any oceans on the surfaces of the bodies while simultaneously equilibrating the isotopic composition of both bodies in the debris cloud surrounding the Earth. This is consistent with the models of Pahlevan and Stevenson (2007) who demonstrated a viable mechanism for equilibrating the oxygen isotopic compositions of the Earth-Moon system via exchange reactions mediated by a moderately long lived disk. Such a disk and its isotopic exchange efficiency would be greatly enhanced by increased water content in the proto-Earth. Depending on the accretional loss of water during the initial accretion of both bodies, there could be several tenths of an Earth mass of water vapor in this cloud. Although most of this water would be lost from the system due to the high temperature of the debris disk, this same high temperature would ensure the equilibration of the silicates in the cloud. If the oceans of the larger proto-Earth dominate the contribution to this disk, then the silicate particles and SiO vapor in the debris cloud will be equilibrated with the Earth before the cloud coalesces into the Moon.

6. Conclusions

Some planetesimals in the early solar system accreted from wide feeding zones and took a long time to heat to sufficiently high temperatures to lose the volatiles and ices they initially contained. Models for the formation of the terrestrial planets suggest that planets grew by the aggregation and growth of proto-planets that themselves grew quickly via runaway accretion from planetesimals in narrow feeding zones. Because these planetesimals had not yet had time to warm to the stage where they would lose a large fraction of their water and volatiles prior to their aggregation into proto-planets, many of those bodies that accreted should have contained reasonably large quantities of ice. The initial composition of the Earth contained more than enough water to form the modern hydrosphere depending on the position of the snowline, the fraction of planetesimals formed via gravitational instabilities or turbulent aggregation and the overall mass of the solar nebula. Even with substantial accretional loses and atmospheric erosion during the series of giant collisions between proto-planets that formed the terrestrial planets, more than enough water should have been available on and within the Earth to account for several modern oceans, especially when one includes the contributions from the decreasing, but continuous infall of comets to modern times.

We do not need to wonder where the Earth's water came from; it clearly arrived with the planetesimals accumulated by the proto-Earth during the accretion process. Instead we should be asking how all of the initial water was lost from the Earth and what the consequences of these possible loss mechanisms are. Modeling the formation of the Earth and the other terrestrial planets from modern (dry) meteoritic matter, even carbonaceous meteorites, is not appropriate, and is certainly inconsistent with results one gets by following a planetesimal from its origin beyond the snowline, into the accreting planet. What effect might such large quantities of water have had on the geochemical differentiation of proto-planets, or of the Earth prior to the Moon forming event? Could a more massive but water filled proto-Earth better account for the properties of the Earth-Moon system during and after the giant collision with a Mars sized body (e.g., Pahlevan and Stevenson, 2007)? No one has yet investigated such possibilities: There is much more work still to do.

7. References

Blum, J. 1990, in Formation of the Stars and Planets, and the Evolution of the Solar System, ed. B. Battrick, (Noordwijk: ESTEC), 87.
Blum, J., and Wurm, G. 2000, Icarus, 143, 138 – 146.
Cameron, A.G.W. and Benz, W. 1991 Icarus, 92, 204–216.
Canup, R.M. and Asphaug, E. 2001 Nature. 412,708–712.
Canup, R.M. 2004 Icarus. 168. 433–456.

Chabot, N.L. and Haack, H. 2006 In Meteorites and the Early Solar System II (D.S. Lauretta & H.Y. McSween, eds.), Univ. Arizona Press, Tucson 747-741.

Chyba, C.F., Owen, T.C., Ip, W.-H. 1994 In Hazards due to comets and asteroids (T. Gehrels, ed.), Univ. of Arizona Press, Tucson 9-58.

Ciesla, F.J. and Charnley, S.B. 2006 In Meteorites and the Early Solar System II (D.S. Lauretta & H.Y. McSween, eds.), Univ. Arizona Press, Tucson 209–230.

Cuzzi J. N. and Weidenschilling S. J., 2006, in Lauretta D. S., McSween H. Y. Jr, eds, Meteorites and the Early Solar System II. Univ. Arizona press, Tucson, p. 353

Das, A. & Srinivasan, G. 2007 LPSC 38 Abstr. #2370.

Desch, S.J. (2008) LPSC 39 Abstr. #1004.

Drake, M. 2005 Meteoritics & Planet. Sci., 40, 519-527.

Drake, M.J. and Righter, K. 2002 Nature, 416, 39-44.

Goldreich, P. & Ward, W. R., 1973, Astrophys. J. 183, 1051–1062.

Hayashi, C., Nakazawa, K., & Nakagawa, Y. 1985, Protostars and Planets II, 1100-1153;

Hayashi, C 1981 Prog. Theoret. Phys. Suppl. 70, 35-53.

Huss, G. R., Rubin, A. E., Grossman, J.N. 2006 In Meteorites and the Early Solar System II (D.S. Lauretta & H.Y. McSween, eds.), Univ. Arizona Press, Tucson 567-586.

Jacobsen, S. 2003 Science, 300, 1513-1514.

Jacobsen S. B. * Remo J. L. Petaev M. I. Sasselov D. D., 2009, LPSC 40 Abstr. # 2054.

Javoy M., 1995, Geophys. Res. Letter, 22, 2219-2222.

Johansen, A., Oishi, J.S., Mac Low, M.M., Klahr, H., Henning, T. and Youdin, A., 2007 Nature 448, 1022 – 1025.

Johansen, A. and Youdin, A., 2007, Ap.J. 662, 627 – 641.

Kokubo, E. and Isa, S., 1995, Icarus 114, 247 – 257.

Kokubo, E. and Isa, S., 1996, Icarus 123, 180 – 191.

Kokubo, E. and Isa, S., 1998, Icarus 131, 171 – 178.

Kokubo, E. and Isa, S., 2000, Icarus 143, 15 – 27.

La Tourette, T. and Wasserburg, G.J. 1998 EPSL 158, 91-108.

Lunine, J. 2006 In Meteorites and the Early Solar System II (D.S. Lauretta & H.Y. McSween, eds.), Univ. Arizona Press, Tucson 309-319.

Nichols, R. 2006 In Meteorites and the Early Solar System II (D.S. Lauretta & H.Y. McSween, eds.), Univ. Arizona Press, Tucson 463-472.

Nuth, J. A. 2008 Earth Moon Planets 102, 435 – 445.

Ormel, C.W., Cuzzi, J.N. and Tielens, A.G.G.M., 2008, Astrophys. J. 679, 1588 – 1610.

Pahlevan, K. and Stevenson, D.J., 2007, EPSL 262, 438 – 449.

Righter,K., Drake, M.J., Scott, E.,. 2006 In Meteorites and the Early Solar System II (D.S. Lauretta & H.Y. McSween, eds.), Univ. Arizona Press, Tucson 803-828.

Righter, K. 2007 Chemie der Erde, 67, 179-200.

Ringwood, A.E., 1979, Origin of the Earth and Moon. Springer-Verlag, New York.

Weidenschilling, S. (1997) Icarus, 127, 290-306.

Wadhwa, M., Amelin, Y., Davis, A.M., Lugmair, G.W., Meyer, B., Gounelle, M., Desch, S.J. 2006 In Protostars and Planets V (B. Reipurth, D. Jewett and K. Keil, eds.), 835-848, Univ. of Arizona Press, Tucson.

Wänke, H., 1981, *Phil. Trans. Roy. Soc. Lond.* A 303, 287–302.

Wetherill, G. and Stewart, G. 1989 Icarus, 77, 330-357.

Wetherill, G. and Stewart, G. 1993 Icarus, 106, 190 - 209.

Wurm, G., Blum, J., and Colwell, J. E. 2001, Icarus, 151, 318 – 321.

Youdin, A. N. & Shu, F. H., 2002, Astrophys. J. 580, 494–505.

Youdin, A. N. & Goodman, J., 2005, Astrophys. J. 620, 459–469.

3

Geomagnetically Induced Currents as Ground Effects of Space Weather

Risto Pirjola
[1]Finnish Meteorological Institute
[2]Natural Resources Canada
[1]Finland
[2]Canada

1. Introduction

"Space Weather" refers to electromagnetic and particle conditions in the near-Earth space. It is controlled by solar activity. The whole space weather chain extending from the Sun to the Earth's surface is very complicated and includes plasma physical processes, in which the interaction of the solar wind with the geomagnetic field plays an essential role. Space weather phenomena statistically follow the eleven-year sunspot cycle but large space weather storms can also occur during sunspot minima. Changes of currents in the Earth's magnetosphere and ionosphere during a space weather storm produce temporal variations of the geomagnetic field, i.e. geomagnetic disturbances and storms. Technological systems, even humans, in space and on the ground may experience adverse effects from space weather (e.g. Lanzerotti et al., 1999).

At the Earth's surface, space weather manifests itself as "Geomagnetically Induced Currents" (GIC) in technological conductor networks, such as electric power transmission grids, oil and gas pipelines, telecommunication cables and railway circuits. GIC observations have a much longer history than the time when the concept of space weather has been used as GIC effects were already found in the first telegraph equipment in the mid-1800's (Boteler et al., 1998; Lanzerotti et al., 1999; Lanzerotti, 2010). Telecommunication systems have suffered from GIC problems several times in the past. Optical fibre cables generally used today are not directly affected by space weather. However, metal wires lying in parallel with fibre cables are used to provide power to repeater stations, and they may be prone to GIC impacts. Trans-oceanic submarine communication cables are a special category regarding GIC since their lengths imply that the end-to-end voltages associated with GIC can be very large.

Buried pipelines may suffer from serious corrosion of the steel due to GIC (e.g. Gummow, 2002). Corrosion is an electrochemical process occurring at points where a current flows from the pipe to the soil. Roughly speaking, a continuous dc current of 1 A for one year causes a loss of about 10 kg of steel. To prevent or minimise corrosion, so-called cathodic protection (CP) systems are used for pipelines. They keep the pipeline in a negative voltage of typically slightly less than 1 V with respect to the soil. Pipe-to-soil voltages associated with GIC may well exceed the CP voltage, thus possibly cancelling and invalidating the protection. Furthermore, control surveys of pipe-to-soil voltages during space weather storms may lead to incorrect data. Pipelines are covered with a highly-resistive coating, whose materials have a much larger resistivity today than in earlier times. But a high resistance also increases pipe-to-soil voltages implying larger harmful currents at possible defects in the coating.

GIC impacts on railways have not been much investigated yet but evidence of anomalies in railway signalling systems due to GIC exists at least in Sweden and Russia (Ptitsyna et al., 2008; Wik et al., 2009; Eroshenko et al., 2010).

Nowadays electric power transmission networks are the most important regarding GIC effects, and the importance continuously increases with the extension of power grids including complex continent-wide interconnections and with the even larger dependence of the society on the availability and reliability of electricity. The frequencies associated with GIC are typically very much lower than the 50/60 Hz ac frequency used in power transmission. Thus, from the viewpoint of a power system, GIC are (quasi-)dc currents. Consequently their presence may lead to half-cycle saturation of transformers, which can result in non-linear behaviour of transformers (e.g. Molinski, 2002; Kappenman, 2007). This further implies large asymmetric exciting currents producing harmonics, unnecessary relay trippings, increased reactive power demands, voltage fluctuations, and possibly even a collapse of the whole power network. Transformers can also be overheated with possible damage. The best-known GIC disturbance is a province-wide blackout in Québec, Canada, for several hours during a large geomagnetic storm in March 1989 (e.g. Bolduc, 2002). A transformer was permanently damaged and had to be replaced in New Jersey, USA, during the same storm (Kappenman & Albertson, 1990).

All this means that research of GIC and space weather is not only relevant and significant regarding space science but important practical applications also exist. As indicated above, today's GIC research is particularly concentrated on power networks, which constitute the main focus in this paper as well. The discussion is limited to space physical and geophysical aspects associated with GIC including the calculation of GIC but neglecting the consideration of engineering details of possible adverse impacts of GIC on networks and their equipment and the discussion of mitigation means against GIC problems.

The shape of the geomagnetic field implies that geomagnetic storms are the most intense and most frequent at high latitudes. So GIC are a special concern in the same areas. However, during major space weather storms, large geomagnetic disturbances may also occur at much lower latitudes, which indicates the possibility of GIC problems there, too. Moreover, GIC magnitudes in a system depend significantly on the network topology, configuration and resistances. GIC values also vary much from site to site in a system being generally large at ends and corners of a network. In addition, the sensitivity of a system to GIC depends on many technical matters and varies from one network to another. Consequently, a GIC value that can be ignored in one system may be hazardous in another. All this shows that GIC issues have to be taken into account in mid- and low-latitude networks, too (e.g. Kappenman, 2003; Trivedi et al., 2007; Bernhardi et al., 2008; Liu et al., 2009a, 2009b).

Finland is located at high latitudes. Consequently, GIC would be a potential problem in the country, and in fact, research of GIC has been carried out as collaboration between Finnish power and pipeline industry and the Finnish Meteorological Institute since the latter part of the 1970's. However, fortunately, GIC have never caused significant problems in Finland

(Elovaara et al., 1992; Elovaara, 2007) whereas the neighbour country, Sweden, has experienced harmful GIC effects several times (Elovaara et al., 1992; Pirjola & Boteler, 2006; Wik et al., 2009). Such a dissimilarity between Sweden and Finland is somewhat surprising but, concerning power systems, it can be explained based on differences in transformer design and specifications in the two countries.

In Section 2 of this paper, we consider research of GIC, which can be done by measurements or by theoretical modelling. Section 3 is devoted to a more detailed discussion of the calculation of GIC in power networks and pipelines, and Section 4 contains concluding remarks.

2. Research of GIC

Research of GIC is highly multidisciplinary since the subjects involved cover items from solar physics to engineering details of the operation of power systems or other networks. We often speak about the "space weather chain" that begins at solar activity, extends via the solar wind and magnetospheric-ionospheric processes to GIC in ground-based systems and to the mitigation of adverse effects of GIC (e.g. Pirjola, 2000; Pirjola et al., 2003). Roughly speaking, GIC studies can be divided into two parts, the first of which refers to the space physical and geophysical investigation of GIC in a network, whereas the second part includes the engineering evaluation of effects of GIC on the system in question as well as the design of techniques for mitigating the harmful impacts. This paper deals with the first part.

The flow of GIC in a network is easy to understand based on Faraday's and Ohm's laws. The geomagnetic field experiences temporal variations during a space weather event. According to Faraday's law, they are accompanied by a geoelectric field, which, based on Ohm's law, drives currents in conductors, i.e. GIC in networks. In theoretical discussions, the determination of GIC in a system is usually divided into two parts or steps. The "geophysical part" and the "engineering part" refer to the modelling of the horizontal geoelectric field at the Earth's surface and to the calculation of GIC in the particular network, respectively (e.g. Pirjola, 2002).

GIC can naturally be studied by making measurements or by theoretical modelling. In practice, the validity of the models always has to be verified by measured data. If necessary, the data may enable adjusting model parameter values. Concerning an appropriate model of the Earth's conductivity in southern Sweden, the adjustment is explicitly demonstrated by Wik et al. (2008).

2.1 Measurements of GIC

The usual place for installing GIC recording equipment in a power network is the earthing lead of a transformer neutral. In the normal situation, this particular lead carries no 50/60 Hz current because the sum of the ac currents in the three phases is equal to zero. Furthermore, measuring GIC in the neutral lead directly gives information about the currents flowing through transformer windings, where they can result in harmful saturation. Recordings of GIC can be performed with a coil around the earthing lead. However, for example in the Finnish 400 kV network, a small shunt resistor is utilised in the lead (e.g. Elovaara et al., 1992). The largest GIC magnitudes measured in Finland and in

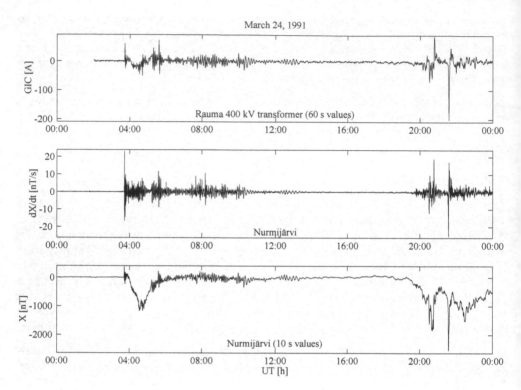

Fig. 1. (*Top*) GIC recorded in the earthing lead of the Rauma 400 kV transformer neutral in southwestern Finland on March 24, 1991, (*Bottom*) north component (X) of the geomagnetic field, and (*Middle*) its time derivative at the Nurmijärvi Geophysical Observatory in southern Finland. The value of 201 A seen in the top plot is the largest measured in the GIC recordings in the Finnish 400 kV system started in 1977 (Pirjola et al., 2003; Pirjola et al., 2005).

Sweden are about 200 A and about 300 A, respectively (Pirjola et al., 2005; Wik et al., 2008). It should be noted that these values give the total GIC, i.e. the sum of GIC in the three phases. The per-phase values are obtained by dividing by three, being thus about 67 A and 100 A. To our knowledge, the Swedish value is the highest ever reported. It occurred during a great geomagnetic storm in April 2000. Figure 1 presents GIC recorded in the earthing lead of the Rauma 400 kV transformer neutral in southwestern Finland on March 24, 1991. It also includes the above-mentioned largest value of about 200 A just before 22 h Universal Time (UT).

The bottom and middle plots of Figure 1 show the recordings of the north component (denoted by X) of the geomagnetic field and its time derivative at the Nurmijärvi Geophysical Observatory located in southern Finland at a distance of about 200 km from Rauma. We see that the GIC curve resembles the derivative, but the behaviour of the geomagnetic field is more different. This would seem to be an expected result because GIC are created based on Faraday's law, in which temporal variations of the magnetic field play a role. However, it is important to note that the geoelectric field and GIC are not directly proportional to the geomagnetic time derivative but the relation is more complicated (see Section 2.2). Pirjola (2010) investigates the relation between geomagnetic variations and the geoelectric field (and GIC) in the case of a simple two-layer Earth. He explicitly shows that a poorly-conducting upper layer above a highly-conducting bottom is favourable to the geoelectric field (and GIC) being proportional to the geomagnetic time derivative at the Earth's surface whereas a thin highly-conducting upper layer above a less-conducting bottom results in the surface geoelectric field (and GIC) proportional to the geomagnetic variation. Trichtchenko & Boteler (2006; 2007) present an example from Canada in which GIC resembles the geomagnetic variation and another example also from Canada in which a correspondence between GIC and the geomagnetic time derivative exists. Watari et al. (2009) demonstrate that in recordings in the Japanese power network GIC show a high correlation with the geomagnetic variation field rather than with the geomagnetic time derivative.

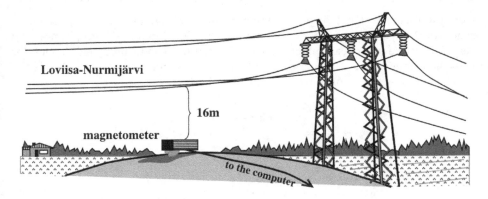

Fig. 2. Measurement of GIC flowing in a power transmission line by using a magnetometer below the line and another magnetometer for reference data further away (Mäkinen, 1993; Viljanen et al., 2009).

Although GIC in transformer neutral earthing leads have a larger practical significance than GIC in transmission lines, both are of the same importance from the scientific point of view. This is a motivation for the measurements of GIC carried out in a transmission line in Finland for some time in the beginning of the 1990's (Mäkinen, 1993; Viljanen et al., 2009). Such measurements are possible to make by using two magnetometers, which behave slowly enough, so that the influence from the ac currents is not experienced. One magnetometer is installed very near the line, as shown in Figure 2. The other is located further away. The former records the sum of the natural geomagnetic field and the field due to GIC in the line whereas the latter only observes the natural field. If the distance of the two magnetometers is not too large (i.e. many tens of kilometres or more), the natural field can be regarded as the same at the two sites. Thus the difference of the magnetometer readings gives the field created by GIC flowing in the line, and by inverting the Biot-Savart law GIC data are obtained. It is naturally necessary to take into account that a transmission line consists of three phase conductors, each of which carries one third of the (total) GIC. The geometries of the phases with respect to the nearby magnetometer are different, which must also be taken into account in the inversion.

A similar two-magnetometer technique has been used successfully to record GIC flowing along the Finnish natural gas pipeline since 1998 (Pulkkinen et al., 2001a; Viljanen et al., 2006). The largest GIC recorded so far is 57 A on October 29, 2003 (Pirjola et al., 2005). The pipe-to-soil voltage is also continuously monitored in the Finnish pipeline as well as in other oil and gas pipelines.

2.2 Modelling of GIC

As pointed out in the beginning of Section 2, GIC modelling is convenient to be carried out in two parts:

1. Determination of the horizontal geoelectric field at the Earth's surface (*"geophysical part"*).
2. Computation of GIC in the network produced by the geoelectric field (*"engineering part"*).

The geophysical part does not depend on the particular network and is thus the same for power networks, pipelines and other conductor systems. The input of the geophysical part consists of knowledge or assumptions about the Earth's conductivity and about the magnetospheric-ionospheric currents or about the geomagnetic variations at the Earth's surface. The solution is based on Maxwell's equations

$$\nabla \cdot \mathbf{E} = \frac{\rho}{\varepsilon_0} \tag{1}$$

$$\nabla \cdot \mathbf{B} = 0 \tag{2}$$

$$\nabla \times \mathbf{E} = -\frac{\partial \mathbf{B}}{\partial t} \tag{3}$$

$$\nabla \times \mathbf{B} = \mu_0 \mathbf{j} + \mu_0 \varepsilon_0 \frac{\partial \mathbf{E}}{\partial t} \tag{4}$$

Maxwell's equations couple the electric field $\mathbf{E}$ [V/m] and the magnetic field $\mathbf{B}$ [Vs/m²] to each other as well as to the charge density ρ [As/m³] and to the current density $\mathbf{j}$ [A/m²]. All these quantities are functions of space $\mathbf{r}$ and time t. The vacuum permeability and the vacuum permittivity are denoted by μ_0 (= $4\pi \cdot 10^{-7}$ Vs/Am) and ε_0 (= $8.854 \cdot 10^{-12}$ As/Vm), respectively. In the form (1)-(4), Maxwell's equations are (microscopically) always valid. However, charges and currents are macroscopically usually divided into different types, and additional fields are introduced. The electric $\mathbf{D}$ field is related to $\mathbf{E}$ and the magnetic $\mathbf{H}$ field is related to $\mathbf{B}$ by constitutive equations, which are usually assumed to be linear and simple as follows

$$\mathbf{D} = \varepsilon \mathbf{E} \tag{5}$$

$$\mathbf{H} = \frac{\mathbf{B}}{\mu} \tag{6}$$

where μ and ε are the permeability and permittivity of the medium. The usual assumption in connection with geoelectromagnetic studies is that $\mu = \mu_0$. Macroscopically, we can write Maxwell's equations similarly to (1)-(4) but μ_0 and ε_0 are replaced by μ and ε, and the charge density $\rho = \rho_{free}$ and the current density $\mathbf{j} = \mathbf{j}_{free}$ should only refer to charges moving freely in the medium. An additional constitutive equation needed in the geophysical part is the Ohm's law that relates $\mathbf{j}_{free}$ and $\mathbf{E}$

$$\mathbf{j}_{free} = \sigma \mathbf{E} \tag{7}$$

where σ is the conductivity of the medium. What is still required for solving the geophysical part are the continuity conditions that enable moving from one medium to another. Usually we utilise the continuity of the tangetial components of the $\mathbf{E}$ and $\mathbf{H}$ fields when moving across a boundary.

Different techniques and models for performing the geophysical part have been investigated for a long time (e.g. Pirjola, 2002, and references therein). An interesting approximate alternative is the Complex Image Method (CIM), in which the currents induced in the conducting Earth are replaced by images of ionospheric currents located in a complex space (Boteler & Pirjola, 1998a; Pirjola and Viljanen, 1998). A crucial parameter in CIM is the complex skin depth $p = p(\omega)$, which depends on the angular frequency ω considered (i.e. we assume a harmonic dependence on the time t given by $\exp(i\omega t)$)

$$p(\omega) = \frac{Z(\omega)}{i\omega\mu_0} \tag{8}$$

where $Z = Z(\omega)$ is the surface impedance at the Earth's surface relating a horizontal electric field component $E_y = E_y(\omega)$ to the perpendicular horizontal magnetic field component $B_x = B_x(\omega)$ (see e.g. Kaufman & Keller, 1981; Pirjola et al., 2009)

$$E_y(\omega) = -\frac{Z(\omega)}{\mu_0} B_x(\omega) \tag{9}$$

It is implicitly required in equation (9) that the (flat) Earth surface is the xy plane of a right-handed Cartesian coordinate system in which the z axis points downwards. In practice, the surface impedance included in equation (9) and especially in equation (8) refers to the plane wave case discussed briefly below.

CIM makes numerical computations much faster than with formulas obtained by a direct forward solution of Maxwell's equations and boundary conditions (Häkkinen & Pirjola, 1986; Pirjola & Häkkinen, 1991). However, it has been proved that, regarding the accuracy and fastness of computations required in GIC applications, the simple plane wave method to be applied locally in different areas covered by the particular network is the best and most practical technique (Viljanen et al., 2004). It should also be noted that, in GIC calculations, it is not necessary to know the spatial details of the geoelectric field exactly because the (geo)voltages driving GIC are obtained by integrating the geoelectric field (see Section 3.1).

In the plane wave method, the geoelectromagnetic disturbance produced by the (primary) magnetospheric-ionospheric currents is a plane wave propagating vertically downwards and the Earth's conductivity structure is layered (with a flat surface) enabling the Earth to be described by a surface impedance (see equation (9)). The contribution to the total geoelectromagnetic disturbance at the Earth's surface from (secondary) currents in the Earth is an upward-propagating reflected wave. This kind of a model is already included in the basic paper of magnetotellurics by Cagniard (1953). It is necessary to emphasise that the frequencies involved in geoelectromagnetic studies are typically in the mHz range and at least below 1 Hz, i.e. so small that the displacement currents (= $\varepsilon \partial E / \partial t$) can practically always be neglected. It means that "geoelectromagnetic plane waves" are actually not "waves", so the terminology generally used in geoelectromagnetics is not completely correct in this respect.

Let us assume now that the Earth is uniform with the conductivity σ and consider a harmonic time dependence with the angular frequency ω. It is easy to show that the horizontal geoelectric field component $E_y = E_y(\omega)$ at the Earth's surface is related to the perpendicular horizontal geomagnetic variation component $B_x = B_x(\omega)$ by the following equation (e.g. Pirjola, 1982)

$$E_y = -\sqrt{\frac{\omega}{\mu_0 \sigma}} e^{i\frac{\pi}{4}} B_x \tag{10}$$

Equation (10) shows that there is a 45-degree ($\pi/4$-radian) phase shift between the geoelectric and geomagnetic fields. We also see that an increase of the angular frequency and a decrease of the Earth's conductivity enhance the geoelectric field with respect to the geomagnetic field.

Noting that $i\omega B_x(\omega)$ is associated with the time derivative of $B_x(t)$, equation (10) can be inverse-Fourier transformed to give the following time domain convolution integral (Cagniard, 1953; Pirjola, 1982)

$$E_y(t) = -\frac{1}{\sqrt{\pi \mu_0 \sigma}} \int_0^\infty \frac{g(t-u)}{\sqrt{u}} du \tag{11}$$

where the time derivative of $B_x(t)$ is denoted by $g(t)$. The derivation of equation (10) makes use of the neglect of the displacement currents, which thus also affects equation (11). If the displacement currents are included the kernel function convolved with $g(t)$ in formula (11) is more complicated containing the Bessel function of the zero order (Pirjola, 1982). Equation (11) is in agreement with the causality, i.e. $E_y(t)$ at the time t only depends on earlier values of $g(t)$. The square root of the lag time u in the denominator means that the influence of a value of $g(t-u)$ on $E_y(t)$ decreases with increasing u. As indicated in Section 2.1, equation (11) shows that the relation of the geoelectric field (and GIC) with the geomagnetic time derivative is not simple, like for example a proportionality, not even in the present plane-wave and uniform-Earth situation.

Similarly to the inverse-Fourier transform of equation (10) leading to (11) in the time domain, we can inverse-Fourier transform equation (9) to get a convolution relation between E_y and B_x in the time domain, or as above in equation (11), perhaps rather between E_y and dB_x/dt.

The engineering part of GIC modelling utilises the horizontal geoelectric field to be provided by the geophysical part and also needs the knowledge of the topology, configuration and all resistance values of the network in question. The computation of GIC can be performed by applying electric circuit theory, i.e. Ohm's and Kirchhoff's laws. Because the frequencies in connection with geoelectromagnetic fields and GIC are very low, a dc treatment is appropriate to the engineering part (at least as the first approximation). Discretely-earthed networks, such as a power system, and continuously-earthed networks, such as a buried pipeline, need to have different calculation techniques. For the former, matrix formulas are available, which enable the computation of GIC between the Earth and the network at the nodes and in the lines between the nodes, whereas the latter can be treated by utilising the distributed-source transmission line (DSTL) theory. These methods are discussed more in Section 3.

Assuming that the horizontal geoelectric field impacting a network is uniform and applying the engineering part techniques, we can easily identify the sites that will most probably experience the largest GIC magnitudes being thus risky for problems, but being also ideal GIC recording sites. With a uniform geoelectric field, we may also simply get a comprehension of the effect of disconnecting or connecting some lines on GIC values in the network. Performing both the geophysical and the engineering parts enables studies of GIC as functions of time at different sites of the system during large space weather storms. Using long-term geomagnetic data recorded at observatories and other magnetometer stations, it is possible to derive statistics of expected GIC values at different sites of a technological network.

3. Calculation of GIC

3.1 Power networks

For investigating the engineering part of GIC modelling in the case of a power system, we consider a network of conductors with N discrete nodes, called stations and earthed by the resistances $R_{e,i}$ ($i = 1,...,N$). Let us assume that the network is impacted by a horizontal geoelectric field E, which implies the flow of geomagnetically induced currents (GIC). Lehtinen and Pirjola (1985) derive a formula for the $N \times 1$ column matrix I_e that includes the currents $I_{e,m}$ ($m = 1,...,N$) called earthing currents or earthing GIC and flowing between the network and the Earth as follows

$$I_e = (1 + Y_n Z_e)^{-1} J_e \qquad (12)$$

The current is defined to be positive when it flows from the network to the Earth and negative when it flows from the Earth to the network. The symbol **1** denotes the $N \times N$ unit identity matrix. The $N \times N$ earthing impedance matrix Z_e and the $N \times N$ network admittance matrix Y_n as well as the $N \times 1$ column matrix J_e are explained below.

The definition of Z_e states that multiplying the earthing current matrix I_e by Z_e gives the voltages between the earthing points and a remote Earth that are related to the flow of the currents $I_{e,m}$ ($m = 1,...,N$). Thus, expressing the voltages by an $N \times 1$ column matrix U, we have

$$U = Z_e I_e \qquad (13)$$

Utilising the reciprocity theorem, Z_e can be shown to be a symmetric matrix. The diagonal elements of Z_e equal the earthing resistances of the stations. If the distances of the stations are large enough, the influence of the current $I_{e,m}$ at one station on the voltages at other stations is negligible, and then the off-diagonal elements of $\mathbf{Z_e}$ are zero (see Pirjola, 2008).

The matrix $\mathbf{Y_n}$ is defined by

$$(i \neq m): Y_{n,im} = -\frac{1}{R_{n,im}} \, , \quad (i = m): Y_{n,im} = \sum_{k=1,k\neq i}^{N} \frac{1}{R_{n,ik}} \tag{14}$$

where $R_{n,im}$ is the resistance of the conductor between stations i and m $(i, m = 1,...,N)$. (If stations i and m are not directly connected by a conductor, $R_{n,im}$ naturally gets an infinite value.) It is seen from equation (14) that $\mathbf{Y_n}$ is symmetric.

The elements $J_{e,m}$ $(m = 1,...,N)$ of the column matrix $\mathbf{J_e}$ are given by

$$J_{e,m} = \sum_{i=1,i\neq m}^{N} \frac{V_{im}}{R_{n,im}} \tag{15}$$

The geovoltage V_{im} is produced by the horizontal geoelectric field $\mathbf{E}$ along the path defined by the conductor line from station i to station m $(i, m = 1,...,N)$, i.e.

$$V_m = \int_i^m \mathbf{E} \cdot d\mathbf{s} \tag{16}$$

Generally, the horizontal geoelectric field is rotational, i.e. the vertical component of the curl of $\mathbf{E}$ is not zero. As seen from equation (3), this is the case when the time derivative of the vertical component of the magnetic field differs from zero. Consequently, the integral in equation (16) is path-dependent. Thus, as indicated, the integration route must follow the conductor between i and m (Boteler and Pirjola, 1998b; Pirjola, 2000). Equations (15) and (16) show that the column matrix $\mathbf{J_e}$ involves the contribution from the geoelectric field $\mathbf{E}$ to equation (12). Note that $\mathbf{I_e}$ and $\mathbf{J_e}$ are equal in the case of perfect earthings, i.e. when $\mathbf{Z_e} = 0$.

Pirjola (2007) presents an alternative, but equivalent, expression for the matrix $\mathbf{I_e}$, which includes the geovoltages more explicitly. It makes use of the total admittance matrix that is defined to be the sum of $\mathbf{Y_n}$ and the inverse of $\mathbf{Z_e}$. However, the application of this alternative expression obviously does not produce any noticeable advantages, for example, in numerical computations.

So far, we have discussed GIC flowing between the network and the Earth. There is also a simple expression for GIC in the conductors between the nodes (e.g. Pirjola, 2007). However, we ignore it and concentrate only on earthing GIC because, in the case of a power system, GIC to (or from) the Earth constitute the harmful currents in transformer windings and they are usually measured (see Section 2.1).

When GIC in a power network are calculated, the three phases are usually treated as one circuit element. The resistance of the element is then one third of that of a single phase, and the GIC flowing in the element is three times the current in a single phase. Moreover, the earthing resistances (which might be called the *total* earthing resistances) are convenient to be assumed to include the actual earthing resistances, the transformer resistances and the resistances of possible reactors or any resistors in the earthing leads of transformer neutrals, with all these resistances connected in series.

As Mäkinen (1993) and Pirjola (2005) present, special modelling techniques are needed for transformers when two different voltage levels are considered in the calculation of GIC in a power network. Moreover, the treatment of autotransformers differs from the case of full two-winding transformers. It is worth mentioning that recent, still unpublished, investigations indicate that the methods and equations described by Mäkinen (1993) and Pirjola (2005) are unnecessarily complicated, though correct.

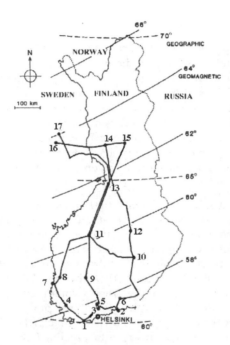

Fig. 3. Finnish 400-kV electric power transmission grid in its configuration valid in October 1978 to November 1979 (Pirjola & Lehtinen, 1985; Pirjola, 2005; Pirjola, 2009). This network can be used as a "test model" for GIC calculation algorithms and programs.

Pirjola (2009) introduces an old version of Finland's 400 kV power network as a "test model" for GIC computation algorithms and programs. The network is shown in Figure 3. It consists of 17 stations and 19 transmission lines, so that it is complex enough to reveal essential features included in GIC computations but not too large to unnecessarily complicate the calculations and analyses. Pirjola (2009) provides the (Cartesian) north and east coordinates of the locations of the stations included in the test model and numbered from 1 to 17. The (total) earthing resistances of the stations as well as the resistances of the transmission lines are also given. Regarding the use of the network as a test model, Pirjola (2009) presents GIC values at all stations and in every transmission line that a uniform (geographically) eastward or a uniform (geographically) northward geoelectric field of 1

V/km creates. Here it should be noted that, by using a linear superposition, these data enable the computation of GIC due to any uniform horizontal geoelectric field impacting the network. For some additional details about the test model, Pirjola (2009) is referred to.

The Rauma 400 kV station discussed in Section 2.1 and especially in Figure 1 was not yet included in the Finnish 400 kV system in October 1978 to November 1979. It is located in the line between stations 4 and 7 quite near station 7. Note that geographic and geomagnetic latitudes are far from being parallel in Finland and that the former are higher. In North America, the geographic latitudes are lower than the geomagnetic since the north geomagnetic pole is on the American side of the geographic pole.

3.2 Pipelines

Pirjola & Lehtinen (1985) present theoretical computations of GIC for the Finnish natural gas pipeline by using the matrix formalism discussed in Section 3.1 and appropriate to a discretely-earthed network, such as a power system. The assumption is included in the treatment by Pirjola & Lehtinen (1985) that the insulating coating of the pipeline is ideal and perfect with a zero conductivity and that the pipeline is earthed at the cathodic protection stations. This approximation should not be considered very good because the large surface area of the pipe makes the pipeline continuously earthed in practice even though the conductivity of the coating material is very small. Furthermore, the CP stations do not constitute normal earthings since the current can only go to the Earth there (to return from the ground to the pipe elsewhere).

Viljanen (1989) presents a GIC study about the Finnish natural gas pipeline based on the simplified assumption that the pipeline is an infinitely long multi-layered cylindrical structure in a homogeneous medium. The model is in agreement with the continuous earthing but it is otherwise too much idealised. According to this model, GIC flowing along the pipeline may reach values of hundreds of amps, which are clearly larger than those measured in Finland (Section 2.1). Although Viljanen (1989) also estimates the effects of a horizontal change of the Earth's conductivity, and of a bend of the pipeline, the treatment is not yet complete for a real pipeline network. A significant improvement in theoretical modelling is shown by Boteler (1997) by incorporating the distributed-source transmission line theory into pipeline-GIC calculations. In fact, the applicability of the DSTL theory is already considered by Boteler and Cookson (1986). An extension is provided by Pulkkinen et al. (2001b) as they also treat branches of a pipeline network.

In the DSTL theory, the pipeline is considered a transmission line containing a series impedance (or resistance due to the dc treatment) Z determined by the properties of the pipeline steel and a parallel admittance Y associated with the resistivity of the coating. The geoelectric field affecting the pipeline forms the distributed source. An important parameter, called the adjustment distance, is the inverse of the propagation constant γ defined by

$$\gamma = \sqrt{ZY} \tag{17}$$

Typical values of the adjustment distance of a real pipeline are tens of kilometres. For the Finnish natural gas pipeline, $Z = 5...9 \cdot 10^{-3}$ Ωkm^{-1} and $Y = 5 \cdot 10^{-2}...0.25$ Ω^{-1} km^{-1} with the exact values depending on the radius of the pipeline at the particular section. Thus, the adjustment distance has values from about 20 km to about 60 km.

Concerning GIC and pipe-to-soil voltages, electrically long pipelines (*length >> adjustment distance*) behave differently than electrically short pipes (*length << adjustment distance*). For a long pipeline, the voltage decays exponentially at a distance comparable to the adjustment distance, when moving from either end of the pipeline towards the centre where it is practically zero. In the same areas near the ends, GIC flows between the pipe and the soil, and in the central parts the current along the pipe is spatially constant. For a short pipeline, the voltage changes linearly along the pipeline, and the current along the pipe is small when the ends are insulated from the Earth. Consequently, central parts of a long pipeline are not critical regarding corrosion problems due to GIC. On the other hand, the pipe-to-soil voltages remain smaller in short pipelines than the end voltages of a long pipeline. Therefore, it might sometimes be reasonable to install insulating flanges in series in a pipeline to break it into shorter sections.

The handling of inhomogeneities associated with a pipeline, such as bends, changes of the pipeline material or of the pipeline size, and branches of the pipeline network, can be accomplished by applying Thévenin's theorem. It expresses an equivalent voltage source and an impedance that describe an external circuit at the terminals of a network section considered. Thus at a pipeline inhomogeneity, Thévenin's voltage and impedance of a section terminating at the inhomogeneity have to be calculated when considering GIC and pipe-to-soil voltages in another section. In this way it is possible to go through a whole pipeline network from section to section. Branches require a special treatment since, for a particular section, Thévenin's components of the other sections are connected in parallel (Pulkkinen et al., 2001b).

4. Conclusion

Geomagnetically induced currents (GIC) are ground effects of space weather, which is associated with a complex chain of phenomena extending from processes in the Sun to GIC in technological networks. In general, GIC are a possible source of problems to the network. Thus research of GIC is practically important, but it also has scientific significance because a ground-based network carrying GIC can be regarded as a huge antenna that collects information from processes in space and within the Earth.

The first GIC observations date back to early telegraph systems more than 150 years ago. Today power systems constitute the most important target of GIC research. The problems in power networks result from half-cycle saturation of transformers created by dc-like GIC. In the worst cases, large areas may experience a blackout due to GIC and transformers can be permanently damaged. The most significant GIC event so far is the blackout in Québec, Canada, for several hours during a large geomagnetic storm in March 1989. Another well-known GIC event caused a blackout in southern Sweden during the so-called "Halloween storm" at the end of October 2003. In 2011 to 2014, an EU-funded project is going on in which GIC in the entire European high-voltage system are considered.

In this paper, we discuss the techniques readily available for calculating GIC values in power networks and pipelines. Future research efforts should be focussed on the application of the methods to data characterising extremely large space weather events, such as the famous Carrington storm in 1859. Such studies would lead to extreme, but realistic, scenarios of GIC magnitudes to be utilised in the estimation of possible problems and in the design of countermeasures.

5. Acknowledgment

The author wishes to thank his colleagues for excellent collaboration in GIC research during many years. Special thanks go to Drs. David Boteler (Canada), Antti Pulkkinen (Finland & USA), Larisa Trichtchenko (Canada), Ari Viljanen (Finland) and Magnus Wik (Sweden) (in alphabetical order). The author also wants to acknowledge the interest and support that Finnish power and pipeline industry have shown to GIC investigations during more than thirty years.

6. References

Bernhardi, E. H.; Cilliers, P. J. & Gaunt, C. T. (2008). Improvement in the modelling of geomagnetically induced currents in southern Africa. *South African Journal of Science*, Vol.104, July/August 2008, pp. 265-272

Bolduc, L. (2002). GIC observations and studies in the Hydro-Québec power system. *Journal of Atmospheric and Solar-Terrestrial Physics*, Vol.64, No.16, pp. 1793-1802

Boteler, D. H. (1997). Distributed-Source Transmission Line Theory for Electromagnetic Induction Studies. *Supplement of the Proceedings of the 12th International Zurich Symposium and Technical Exhibition on Electromagnetic Compatibility*, Zürich, Switzerland, February 18-20, 1997, OE7, pp. 401–408

Boteler, D. & Cookson, M. J. (1986). Telluric currents and their effects on pipelines in the Cook Strait region of New Zealand. *Materials Performance*, March 1986, pp. 27-32

Boteler, D. H. & Pirjola, R. J. (1998a). The complex-image method for calculating the magnetic and electric fields produced at the surface of the Earth by the auroral electrojet. *Geophysical Journal International*, Vol.132, No.1, pp. 31-40

Boteler, D. H. & Pirjola, R. J. (1998b). Modelling Geomagnetically Induced Currents produced by Realistic and Uniform Electric Fields. *IEEE Transactions on Power Delivery*, Vol.13, No.4, pp. 1303-1308

Boteler, D. H.; Pirjola, R. J. & Nevanlinna, H. (1998). The effects of geomagnetic disturbances on electrical systems at the earth's surface. *Advances in Space Research*, Vol.22, No.1, pp. 17-27

Cagniard, L. (1953). Basic theory of the magnetotelluric method of geophysical prospecting. *Geophysics*, Vol.18, pp. 605–635

Elovaara, J. (2007). Finnish experiences with grid effects of GIC's. In: *Space Weather, Research towards Applications in Europe*, J. Lilensten (Ed.), Astrophysics and Space Science Library, 344, ESA, COST 724, Springer, Chapter 5.4, pp. 311-326

Elovaara, J.; Lindblad, P.; Viljanen, A.; Mäkinen, T.; Pirjola, R.; Larsson, S. & Kielén, B. (1992). Geomagnetically induced currents in the Nordic power system and their effects on equipment, control, protection and operation. *CIGRÉ Paper* No. 36–301

(CIGRÉ = International Conference on Large High Voltage Electric Systems), *CIGRÉ General Session 1992*, Paris, France, August 31 – September 5, 1992, No. 36–301, 10 pp.

Eroshenko, E. A.; Belov, A. V.; Boteler, D.; Gaidash, S. P.; Lobkov, S. L.; Pirjola, R. & Trichtchenko, L. (2010). Effects of strong geomagnetic storms on Northern railways in Russia. *Advances in Space Research*, Vol.46, No.9, pp. 1102-1110

Gummow, R. A. (2002). GIC effects on pipeline corrosion and corrosion control systems. *Journal of Atmospheric and Solar-Terrestrial Physics*, Vol.64, No.16, pp. 1755-1764

Häkkinen, L. & Pirjola, R. (1986). Calculation of electric and magnetic fields due to an electrojet current system above a layered earth. *Geophysica*, Vol.22, Nos.1-2, pp. 31-44

Kappenman, J. G. (2003). Storm sudden commencement events and the associated geomagnetically induced current risks to ground-based systems at low-latitude and midlatitude locations. *Space Weather*, Vol.1, No.3, 1016, doi:10.1029/2003SW000009, 16 pp.

Kappenman, J. G. (2007). Geomagnetic Disturbances and Impacts upon Power System Operation. In: *The Electric Power Engineering Handbook*, L. L. Grigsby (Ed.), CRC Press/IEEE Press, 2nd Edition, Chapter 16, pp. 16-1 - 16-22

Kappenman, J. G. & Albertson, V. D. (1990). Bracing for the geomagnetic storms. *IEEE Spectrum*, March 1990, pp. 27-33

Kaufman, A. A. & G. V. Keller, G. V. (1981). *The Magnetotelluric Sounding Method*, Methods in Geochemistry and Geophysics, 15, Elsevier Scientific Publishing Company, 595 pp.

Lanzerotti, L. J. (2010). Using the Guide of History, *Space Weather*, Vol.8, S03004, doi:10.1029/2010SW000579, 2 pp.

Lanzerotti, L. J.; Thomson, D. J. & Maclennan, C. G. (1999). Engineering issues in space weather, In: *Modern Radio Science 1999*, M. A. Stuchly (Ed.), International Union of Radio Science (URSI), Oxford University Press, pp. 25-50

Lehtinen, M. & Pirjola, R. (1985). Currents produced in earthed conductor networks by geomagnetically-induced electric fields. *Annales Geophysicae*, Vol.3, No.4, pp. 479-484

Liu, C.-M.; Liu, L.-G. & Pirjola, R. (2009a). Geomagnetically Induced Currents in the High-Voltage Power Grid in China. *IEEE Transactions on Power Delivery*, Vol.24, No.4, pp. 2368-2374

Liu, C.-M.; Liu, L.-G.; Pirjola, R. & Wang, Z.-Z. (2009b). Calculation of geomagnetically induced currents in mid- to low-latitude power grids based on the plane wave method: A preliminary case study. *Space Weather*, Vol.7, No.4, S04005, doi: 10.1029/2008SW000439, 9 pp.

Mäkinen, T. (1993). Geomagnetically induced currents in the Finnish power transmission system. *Finnish Meteorological Institute, Geophysical Publications*, No.32, Helsinki, Finland, 101 pp.

Molinski, T. S. (2002). Why utilities respect geomagnetically induced currents. *Journal of Atmospheric and Solar-Terrestrial Physics*, Vol.64, No.16, pp. 1765-1778

Pirjola, R. (1982). Electromagnetic induction in the earth by a plane wave or by fields of line currents harmonic in time and space. *Geophysica*, Vol.18, Nos.1-2, pp. 1-161

Pirjola, R. (2000). Geomagnetically Induced Currents During Magnetic Storms. *IEEE Transactions on Plasma Science*, Vol.28, No.6, pp. 1867-1873

Pirjola, R. (2002). Review on the calculation of surface electric and magnetic fields and of geomagnetically induced currents in ground-based technological systems. *Surveys in Geophysics*, Vol.23, No.1, pp. 71-90

Pirjola, R. (2005). Effects of space weather on high-latitude ground systems. *Advances in Space Research*, Vol.36, No.12, doi: 10.1016/j.asr.2003.04.074, pp. 2231-2240

Pirjola, R. (2007). Calculation of geomagnetically induced currents (GIC) in a high-voltage electric power transmission system and estimation of effects of overhead shield wires on GIC modelling. *Journal of Atmospheric and Solar-Terrestrial Physics*, Vol.69, No.12, pp. 1305-1311

Pirjola, R. (2008). Effects of interactions between stations on the calculation of geomagnetically induced currents in an electric power transmission system. *Earth, Planets and Space*, Vol.60, No.7, pp. 743-751

Pirjola, R. (2009). Properties of matrices included in the calculation of geomagnetically induced currents (GICs) in power systems and introduction of a test model for GIC computation algorithms. *Earth, Planets and Space*, Vol.61, No.2, pp. 263-272

Pirjola, R. (2010). Derivation of characteristics of the relation between geomagnetic and geoelectric variation fields from the surface impedance for a two-layer earth. *Earth, Planes and Space*, Vol.62, No.3, pp. 287-295

Pirjola, R. J. & Boteler, D. H. (2006). Geomagnetically induced currents in European high-voltage power systems. *CD ROM Proceedings of the Canadian Conference on Electrical and Computer Engineering (CCECE), IEEE Ottawa*, Ottawa, Canada, May 7-10, 2006, Paper 820, 4 pp.

Pirjola, R. J. & Häkkinen, L. V. T. (1991). Electromagnetic Field Caused by an Auroral Electrojet Current System Model. In: *Environmental and Space Electromagnetics*, H. Kikuchi (Ed.), Springer-Verlag, Tokyo, Printed in Hong Kong, Chapter 6.5, pp. 288-298

Pirjola, R. & Lehtinen, M. (1985). Currents produced in the Finnish 400 kV power transmission grid and in the Finnish natural gas pipeline by geomagnetically-induced electric fields. *Annales Geophysicae*, Vol.3, No.4, pp. 485-491

Pirjola, R. & Viljanen, A. (1998). Complex image method for calculating electric and magnetic fields produced by an auroral electrojet of finite length. *Annales Geophysicae*, Vol.16, No.11, pp. 1434-1444

Pirjola, R.; Boteler, D. & Trichtchenko, L. (2009). Ground effects of space weather investigated by the surface impedance. *Earth, Planets and Space*, Vol.61, No.2, pp. 249-261.

Pirjola, R.; Pulkkinen, A. & Viljanen, A. (2003), Studies of space weather effects on the Finnish natural gas pipeline and on the Finnish high-voltage power system, *Advances in Space Research*, Vol.31, No.4, pp. 795-805

Pirjola, R.; Kauristie, K.; Lappalainen, H.; Viljanen, A. & Pulkkinen, A. (2005). Space weather risk. *Space Weather*, Vol.3, No.2, S02A02, doi: 10.1029/2004SW000112, 11 pp.

Ptitsyna, N. G.; Kasinskii, V. V.; Villoresi, G.; Lyahov, N. N.; Dorman, L. I. & Iucci, N. (2008). Geomagnetic effects on mid-latitude railways: A statistical study of anomalies in the operation of signaling and train control equipment on the East-Siberian Railway. *Advances in Space Research*, Vol.42, No.9, pp. 1510-1514

Pulkkinen, A.; Viljanen, A.; Pajunpää, K. & Pirjola, R. (2001a). Recordings and occurrence of geomagnetically induced currents in the Finnish natural gas pipeline network. *Journal of Applied Geophysics*, Vol. 48, No.4, pp. 219-231

Pulkkinen, A.; Pirjola, R.; Boteler, D.; Viljanen, A. & Yegorov, I. (2001b). Modelling of space weather effects on pipelines. *Journal of Applied Geophysics*, Vol.48, No.4, pp. 233-256

Trichtchenko, L. & Boteler, D. H. (2006). Response of power systems to the temporal characteristics of geomagnetic storms. *CD ROM Proceedings of the Canadian Conference on Electrical and Computer Engineering (CCECE), IEEE Ottawa*, Ottawa, Canada, May 7-10, 2006, Paper 387, 4 pp.

Trichtchenko, L. & Boteler, D. H. (2007). Effects of recent geomagnetic storms on power systems, *Proceedings of the 7-th International Symposium on Electromagnetic Compatibility and Electromagnetic Ecology*, Saint-Petersburg, Russia, June 26-29, 2007, pp. 265-268

Trivedi, N. B.; Vitorello, Í.; Kabata, W.; Dutra, S. L. G.; Padilha, A. L.; Bologna, M. S.; de Pádua, M. B.; Soares, A. P.; Luz, G. S.; de A. Pinto, F.; Pirjola, R. & Viljanen, A. (2007). Geomagnetically induced currents in an electric power transmission system at low latitudes in Brazil: A case study. *Space Weather*, Vol.5, No.4, S04004, doi: 10.1029/2006SW000282, 10 pp.

Viljanen, A. (1989). Geomagnetically Induced Currents in the Finnish Natural Gas Pipeline. *Geophysica*, Vol.25, Nos.1&2, pp. 135-159

Viljanen, A. T., Pirjola, R. J.; Pajunpää, K. M. & Pulkkinen, A. A. (2009). Measurements of geomagnetically induced currents by using two magnetometers. *Proceedings of the 8-th International Symposium on Electromagnetic Compatibility and Electromagnetic Ecology*, Saint-Petersburg, Russia, June 16-19, 2009, pp. 227-230

Viljanen, A.; Pulkkinen, A.; Amm, O.; Pirjola, R.; Korja, T. & BEAR Working Group (2004). Fast computation of the geoelectric field using the method of elementary current systems and planar Earth models. *Annales Geophysicae*, Vol.22, No.1, pp. 101-113

Viljanen, A.; Pulkkinen, A.; Pirjola, R.; Pajunpää, K.; Posio, P. & Koistinen, A. (2006). Recordings of geomagnetically induced currents and a nowcasting service of the Finnish natural gas pipeline system. *Space Weather*, Vol.4, No.10, S10004, doi: 10.1029/2006SW000234, 9 pp.

Watari, S.; Kunitake, M.; Kitamura, K.; Hori, T.; Kikuchi, T.; Shiokawa, K.; Nishitani, N.; Kataoka, R.; Kamide, Y.; Aso, T.; Watanabe, Y. & Tsuneta, Y. (2009). Measurements of geomagnetically induced current in a power grid in Hokkaido, Japan. *Space Weather*, Vol.7, No.3, S03002, doi: 10.1029/2008SW000417, 11 pp.

Wik, M.; Pirjola, R.; Lundstedt, H.; Viljanen, A.; Wintoft, P. & Pulkkinen, A. (2009). Space weather events in July 1982 and October 2003 and the effects of geomagnetically induced currents on Swedish technical systems. *Annales Geophysicae*, Vol.27, No.4, pp. 1775-1787

Wik, M.; Viljanen, A.; Pirjola, R.; Pulkkinen, A.; Wintoft, P. & Lundstedt, H. (2008). Calculation of geomagnetically induced currents in the 400 kV power grid in southern Sweden. *Space Weather*, Vol.6, No.7, S07005, doi: 10.1029/2007SW000343, 11 pp.

Part 3

Planetary Science

4

OrbFit Impact Solutions for Asteroids (99942) Apophis and (144898) 2004 VD17

Wlodarczyk Ireneusz

Chorzow Astronomical Observatory; Rozdrazew Astronomical Observatory
Poland

1. Introduction

The best systems with the exact impact solutions for dangerous asteroids are presented by the JPL Sentry System: http://neo.jpl.nasa.gov/risk/ and by the NEODyS CLOMON2: http://newton.dm.unipi.it/neodys/index.php?pc=4.1

From many years on the top of these lists were two asteroids: (99942) Apophis (is still up now, October 2011) and (144898) 2004 VD17 – now is removed from the list of the dangerous asteroids. Thanks to the courtesy of those who made free available OrbFit software and its source code at: http://adams.dm.unipi.it/~orbmaint/orbfit/

It is now possible to compute individually dates of possible impacts of selected dangerous asteroids or the energy of impact and others impact factors. In this respect we investigated the motion of these recently discovered minor planets: (99942) Apophis and (144898) 2004 VD17 - the most dangerous for the Earth, according to the Impact Risk Page of NASA: http://neo.jpl.nasa.gov/risk/.

To compute exact impact solutions of asteroids it is necessary to include some additional small effects on the asteroid's motion. The inluence of relativistic effects, the perturbing massive asteroids, the Yarkovsky/YORP effects, solar radiation pressure, different ephemeris of the Solar System were investigated. To compute gravitational forces perturbing the motion of (99942) Apophis and (144898) 2004 VD17 from different massive asteroids, the free software Solex from A. Vitagliano was used: http://chemistry.unina.it/~alvitagl/solex/.

SOLEX computes positions of the Solar System bodies by a method which is entirely based on the numerical integration of the Newton equation of motions (Vitagliano, A. 1997). With the use of Solex it was possible to compute all close approaches between (99942) Apophis and (144898) 2004 VD17 with all nearly 140000 numbering asteroids. Similar work with (15) Eunomia using Solex was done by Vitagliano and Stoss (2006).

Selected orbit solutions for (99942) Apophis and (144898) 2004 VD17 were presented during Meeting on Asteroids and Comets in Europe - May 12-14, 2006 in Vienna, Austria. At that time the new version of OrbFit (3.3.2) was released and gave better results of computations of impact probability mainly with the use of non linear monitoring and multi ple solutions method (Milani et al., 2002, Milani et al., 2005a and Milani et al.,

2005b). The main goal of our work was to compare our results generated by OrbFit with the results presented by CLOMON2 system which uses the same OrbFit software and with the results of JPL NASA SENTRY. The second purpose was to prove how differently small effects in motion of asteroid change impact solutions. It was possible thanks to public available source code of the OrbFit software. The orbital uncertainty of an asteroid is viewed as a cloud of possible orbits centered on the nominal solution, where density is greatest. This is represented by the multivariate Gaussian probability density and the use of this probability density relies on the assumption that the observational errors are Gaussian (Milani et al., 2002). Now, August 2011, we have new version of the OrbFit software, v.4.2, implementing the new error model based upon Chesley, Baer and Monet (2010).

2. Some impact solutions for (99942) apophis

2.1 The Influence of *sigma_LOV* and radar observation

The orbital elements of (99942) Apophis in Tab. 1 were computed by the author using all 1007 observations up to this date (Sep. 14th, 2006) and software OrbFit where M - mean anomaly, a - semimajor axis, e - eccentricity, ω_{2000} – argument of perihelion, Ω_{2000} - longitude of the ascending node, i_{2000} - inclination of the orbit. These orbital elements are referred to the *J2000* equator and equinox.

$M[deg]$	$a[AU]$	e	$\omega_{2000}[deg]$	$\Omega_{2000}[deg]$	$i_{2000}[deg]$
333.507245	0.92226793	0.19105946	126.393030	204.460151	3.331317

Table 1. (99942) Apophis: orbital elements. 1007 observations from 885 days (2004/03/15.11 - 2006/08/16.63), *rms*=0.302". Nominal orbit: epoch 2006 Jun. 14.0.

Fig. 1 presents the orbit of (99942) Apophis projected to the ecliptic plane, where *x*-axis is directed to vernal equinox. The dotted lines indicate the part of the orbit below the ecliptic plane. It is clearly seen that orbit of this asteroid crosses the orbit of the Earth and approaches that of Venus. The influence of the radar observations in computations of impact solutions for (99942) Apophis were performed using all observations available before date of MACE 2006. There were 987 optical observations (of which 6 were rejected as outliers) from 2004/03/15.108 to 2006/03/26.509, and also seven radar data points on 2005/01/27, 2005/01/29, 2005/01/31 and 2005/08/07.

Tab. 2 lists impact solutions for (99942) Apophis computed by the author for these settings: multiple solutions, use scaling, line of variation (LOV) with the largest eigenvalue (Milani et al. 2002) in comparison with these published at NEODYS CLOMON2 site: http://newton.dm.unipi.it/neodys/index.php?pc=1.1.2&n=99942 and at the NASA SENTRY site: http://neo.jpl.nasa.gov/risk/a99942.html.

The software OrbFit ver. 3.3.1 for UNIX was used. In this impact table everywhere weighing of observations was as CLOMON2. In Tab. 2 *date* is a calendar day for the potential impact; *dist.[RE]* – minimum distance, the lateral distance from LOV (line of variation, which represent the central axis of the asteroid's elongated uncertainty region); *impact probability* - computed with a Gaussian bidimensional probability density; *IW* –

computed solutions by author of this paper; *nr* denotes solution without radar observations and σ equal to *sigma_LOV* - approximate location along the LOV in sigma space; values of sigma are usually in the interval [-3,3] which represent 99.7 % probability of occurrence of real asteroid in this confidence region (Milani et al. 2002). The impact probability is not reported if the computed value is less than 1E-11. The presented σ are only the input data in OrbFit software, not the real σ - positive or negative, along the LOV. For example σ = 3 denotes that the real σ is between -3 and +3. For different setting of σ value we observe slightly different impact solutions mainly in the date of possible impact. The differences between the results from the NEODyS (CLOMON2) and the JPL NASA (SENTRY) are evident because they are independent systems as state at: http://neo.jpl.nasa.gov/risk/doc/sentry_faq.html. For example impact probabilities different by a factor of ten or so are not extraordinary.

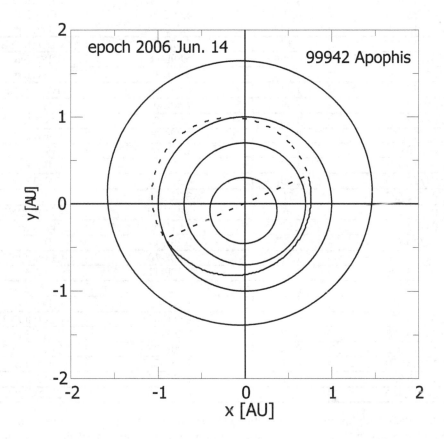

Fig. 1. The orbit of (99942) Apophis projected to the ecliptic plane, where *x*-axis is directed to vernal equinox.

date	dist[RE]	impact probability	source
2035/04/14.131	1.36	1.82E-05	IW, 3 σ, nr
2036/04/13.371	1.15	1.61E-04	CLOMON2
2036/04/13.370	0.53	1.60E-04	SENTRY
2036/04/13.371	1.14	3.46E-05	IW, 3 σ
2036/04/13.371	1.15	1.02E-04	IW, 3 σ, nr
2036/04/13.371	1.14	3.46E-05	IW, 6 σ,
2037/04/13.644	1.36	1.96E-07	CLOMON2
2037/04/13.640	0.63	2.00E-07	SENTRY
2037/04/13.644	1.36	2.46E-05	IW, 3 σ, nr
2037/04/13.644	1.36	1.45E-08	IW, 6 σ
2038/04/13.659	1.73	1.59E-10	CLOMON2
2040/04/13.135	1.65	8.17E-09	IW, 3 σ
2040/04/13.173	1.11	3.40E-08	IW, 3 σ, nr
2042/04/13.726	1.80	3.62E-07	CLOMON2
2042/04/13.710	0.99	4.60E-07	SENTRY
2042/04/13.719	1.38	9.29E-08	IW, 3 σ
2044/04/13.297	2.10	2.57E-07	CLOMON2
2044/04/13.296	2.08	5.89E-08	IW, 3 σ
2044/04/13.264	1.79	6.23E-11	IW, 6 σ
2046/04/13.797	1.98	3.84E-08	IW, 3 σ, nr
2053/04/12.913	1.39	1.80E-07	IW, 3 σ
2054/04/13.401	1.46	6.95E-09	CLOMON2
2054/04/13.400	0.60	7.20E-09	SENTRY
2054/04/13.403	1.27	1.14E-06	IW, 3 σ, nr
2054/04/13.404	1.30	4.79E-10	IW, 6 σ
2055/04/13.730	1.25	4.33E-07	IW, 3 σ, nr
2056/04/12.867	0.70	3.71E-08	IW, 3 σ, nr
2059/04/13.954	2.08	4.31E-10	CLOMON2
2059/04/13.954	2.07	4.81E-08	IW, 3 σ, nr
2059/04/13.953	2.07	3.48E-11	IW, 6 σ
2063/04/13.796	1.30	1.80E-10	CLOMON2
2063/04/13.795	1.26	1.18E-11	IW, 6 σ
2068/04/12.631	0.69	1.77E-06	IW, 3 σ
2068/04/12.631	0.26	1.04E-06	IW, 6 σ
2069/04/13.078	0.97	2.58E-07	CLOMON2
2069/04/13.078	0.99	2.46E-07	IW, 3 σ
2069/10/15.972	0.48	1.20E-07	CLOMON2
2069/10/15.970	0.41	2.55E-07	IW, 3 σ
2069/10/15.970	0.27	2.32E-07	IW, 6 σ
2078/04/13.442	1.93	5.23E-09	CLOMON2

Table 2. (99942) Apophis: influence of different *sigma_LOV* and radar observations for computed impact solutions. Note nr means solution without radar observations. SENTRY and CLOMON2 are systems of impact risk computing of the JPL NASA and the NEODYS, respectively. IW are results by the author.

From results in Tab. 2 we can see that we must include radar observations in computations of impact solutions for (99942) Apophis. Without radar observations we have not impact solutions in 2042, 2044, 2053 and beyond 2063 year. Instead we have mistaken dates of possible impacts in 2035, 2046, 2055 and 2056 years. The usefulness of radar observations is presented e.g. in the paper of Yeomans et. al. (1987). No impact solutions for $\sigma=1$ were found. Time of computations of single solution was about 3 hrs with 1.7 MHz processor.

2.2 (99942) Apophis: Approaching asteroids

To compute exactly impact solutions for (99942) Apophis it is necessary to include gravitational perturbations of approaching massive asteroids. Usually SENTRY include 3 massive asteroids: (1) Ceres, (2) Pallas and (4) Vesta, CLOMMON2 - as SENTRY or 4 asteroids: (1) Ceres, (2) Pallas, (4) Vesta and (10) Hygiea. Using the software Solex ver. 9.0 we have investigated all close approaches of about 140,000 numbered asteroids known in Sept. 2006 with (99942) Apophis within 0.2 AU till 2100 year. We have found 4 asteroids with several close approaches to (99942) Apophis: (433) Eros, (887) Alinda, (1685) Toro and (1866) Sisyphus. These selected asteroids together with the 4 massive ones (Ceres, Pallas, Vesta and Hygiea) were included to equations of motion of (99942) Apophis. The computations of influence of gravitational perturbations of these asteroids for the motion of (99942) Apophis were performed using software OrbFit ver. 3.3.1. The masses of asteroids were taken from Michalak (2001) and from Solex as computed by A. Vitagliano. First of all we must include Ceres in our gravitational model which has about 30 % of the mass of the main belt asteroids and the asteroids which have the closest approaches with (99942) Apophis. All results in Tab. 3 are computed using the JPL Planetary and Lunar Ephemerides DE405 and relativistic effects.

The suitable results in Tab. 3 were computed based on 996 optical observations of which 5 are rejected as outliers from 2004/03/15.108 to 2006/07/27.614, and also on seven radar data points on 2005/01/27, 2005/01/29, 2005/01/31, 2005/08/07 and 2006/05/06.

In Tab. 3 and in all others SENTRY denotes the results from the JPL NASA and CLOMON2 from the NEODYS site. The author results are: *IW-a*: no perturbing asteroids; *IW-b*: Ceres and 4 close approaching asteroids to (99942) Apophis: Eros, Alinda, Toro, Sisyphus; *IW-c*: 4 perturbing asteroids: Ceres, Pallas, Vesta, Hygiea; *IW-d*: 5 perturbing asteroids: Ceres, Pallas, Vesta, Hygiea and approaching asteroid Eros; *IW-e*: 3 perturbing asteroids: Ceres, Pallas and Vesta.

From Table 3 we can see that there is significant role of massive asteroids in motion of Apophis, specially after 2042. Some impact solutions does not exist in given year. For example, in April, 2069 there are only impact solutions with additional perturbing effect from together: Ceres, Eros, Alinda, Toro, Sisyphus and the second solution with perturbations from Ceres, Pallas, Vesta and Hygiea.

Fig. 2 shows the changes of differences in mean anomaly between asteroid (99942) Apophis on nominal orbits for different cases. In Fig. 2(a) there are differences in mean anomaly between (99942) Apophis with and no relativistic effects included. Fig. 2(b) presents differences in mean anomaly of (99942) Apophis between orbits computed without perturbing asteroids and with perturbation from: 1 - Ceres, Pallas and Vesta, 2 - Ceres, Pallas, Vesta, and Hygiea and 3 - Ceres, Pallas, Vesta, Hygiea and Eros. It is clear from Fig.

2(a) that a relativistic effects play a great role in motion of asteroid - over 30 degs difference in mean anomaly between asteroids with and no these effects in the next 100 years. However in Fig. 2(b) the infuence of close approaching asteroids is evident but these effects are several times smaller than the relativistic effects. The rapidly changes in differences in mean anomaly in Fig. 2 are connected with the close approaches of (99942) Apophis with the Earth in the years: 2029 (0.00025 AU) and 2057 (0.022 AU) for the nominal orbits. Hence chaoticity of the motion of the asteroid appears (Wlodarczyk, 2001). The infuence of number of perturbing asteroids on impact solutions for (99942) Apophis lists Tab. 3.

date	dist[RE]	impact probability	source
2036/04/13.370	0.53	2.20E-05	SENTRY
2036/04/13.371	1.15	2.40E-04	CLOMON2
2036/04/13.371	1.15	2.40E-04	IW-a
2036/04/13.371	1.15	2.12E-05	IW-b
2036/04/13.371	1.15	2.39E-04	IW-c
2036/04/13.371	1.15	2.12E-05	IW-d
2036/04/13.371	1.15	2.39E-05	IW-e
2037/04/13.640	0.63	8.5E-08	SENTRY
2042/04/13.715	2.06	6.59E-08	CLOMON2
2042/04/13.718	1.37	6.61E-08	IW-a
2042/04/13.717	1.40	6.09E-08	IW-b
2042/04/13.717	1.38	6.73E-08	IW-c
2042/04/13.717	1.40	6.11E-08	IW-d
2042/04/13.717	1.38	6.73E-08	IW-b
2044/04/13.296	2.09	4.07E-08	CLOMON2
2044/04/13.298	2.13	3.89E-08	IW-a
2044/04/13.294	2.11	3.70E-08	IW-b
2044/04/13.298	2.13	3.45E-08	IW-d
2053/04/12.913	1.39	1.27E-07	CLOMON2
2054/04/13.400	0.59	2.70E-09	SENTRY
2068/04/12.630	0.62	1.79E-06	IW-a
2068/04/12.630	0.52	1.63E-06	IW-c
2068/04/12.633	0.48	8.19E-07	IW-d
2068/04/12.631	0.37	1.03E-06	IW-e
2069/04/13.079	2.00	4.43E-07	CLOMON2
2069/04/13.078	0.97	5.51E-07	IW-b
2069/04/13.079	0.96	4.72E-07	IW-c
2069/10/15.596	0.62	1.02E-07	CLOMON2
2069/10/15.972	0.49	2.63E-07	CLOMON2
2069/10/15.970	0.38	4.70E-07	IW-b
2069/10/15.972	0.59	2.90E-07	IW-c
2069/10/15.971	0.56	4.21E-07	IW-d
2077/04/13.166	1.79	4.33E-08	CLOMON2

Table 3. (99942) Apophis: influence of approaching asteroids for computed impact solutions.

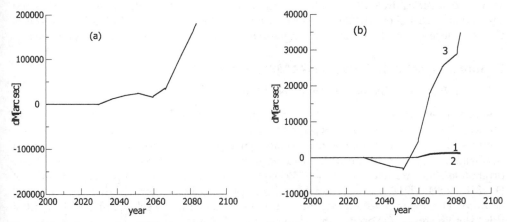

Fig. 2. (99942) Apophis. Differences in mean anomaly between nominal orbits from different solutions: *(a)* - 4 perturbing asteroids: relativistic/non relativistic effects included; *(b)* - different number of perturbing asteroids (see text).

2.3 (99942) Apophis: The JPL Ephemerides

The question was appeared how the model of the Solar System used influences for the impact solutions of (99942) Apophis. Generally the JPL Planetary and Lunar Ephemerides DE203, DE405 or DE406 (SENTRY), DE405 (CLOMON2), DE406 (some in this paper) or DE405/WAW (Sitarski, 2002) were used. DE405 ephemerides (includes both nutations and librations) are computed for time span JED 2305424.50 (1599 DEC 09) to JED 2525008.50 (2201 FEB 20). DE406 is the new "JPL Long Ephemeris" (includes neither nutations nor librations). They works for time span JED 0624976.50 (-3001 FEB 04) to 2816912.50 (+3000 MAY 06). This is the same ephemeris as DE405, though the accuracy of the interpolating polynomials has been lessened. Using OrbFit software v.3.3.2 for Linux and 994 optical observations of (99942) Apophis from 2004/03/15.108 to 2006/06/02.602, and also on seven radar data points on 2005/01/27, 2005/01/29, 2005/01/31, 2005/08/07 and 2006/05/06 we have found some impact results for different planetary ephemerides.

Date	Dist. [RE]	Author	JPL
2036/04/13.371	1.15	CLOMON2	(DE405)
2036/04/13.371	1.15	IW	DE405
2036/04/13.371	1.15	IW	DE406
2042/04/13.720	1.41	CLOMON2	(DE405)
2042/04/13.718	1.37	IW	DE405
2042/04/13.718	1.37	IW	DE406
2044/04/13.295	2.09	CLOMON2	(DE405)
2044/04/13.295	2.08	IW	DE405
2044/04/13.295	2.05	IW	DE406

Table 4. (99942) Apophis: Influence of the JPL Ephemerides on impact solutions.

As we can see in Tab. 4 the results for the JPL ephemerides DE405 and DE406 are almost the same. However Andrea Milani in his e-mail on Juni, 6-th, 2006 wrote: "A particularly good result (using DE405 or DE406), given the strong instability of these solutions, as a result of the very close approach in 2029. My congratulations for your very accurate computations."

3. Some impact solutions for (144898) 2004 VD17

3.1 The influence of *sigma_LOV* and weighting

The orbital elements of (144898) 2004 VD17 presented in Tab. 5 were computed using all known observations up to 14th Sept., 2006 by the author with the software OrbFit 3.3.2 for Linux where M - mean anomaly, a - semimajor axis, e - eccentricity, ω_{2000} - argument of perihelion, Ω_{2000} - longitude of the ascending node, i_{2000} - inclination of the orbit. These orbital elements are referred to the *J2000* equator and equinox. Fig. 3 presents the orbit of (144898) 2004 VD17 projected to the ecliptic plane, where x-axis is directed to vernal equinox. The dotted lines indicate the part of the orbit below the ecliptic plane. The orbit of this asteroid crosses the orbit of the Earth and that of Venus.

$M[deg]$	$a[AU]$	e	$\omega_{2000}[deg]$	$\Omega_{2000}[deg]$	$i_{2000}[deg]$
340.212924	1.5082009	0.58866739	90.686443	224.242137	4.223018

Table 5. (144898) 2004 VD17: orbital elements. 933 observations from 1553 days (2002/02/16.46 - 2006/05/24.10), rms=0.351". Nominal orbit: epoch 2006 Jun. 14.0.

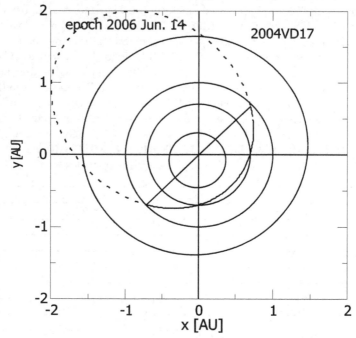

Fig. 3. The orbit of (144898) 2004 VD17 projected to the ecliptic plane, where x-axis is directed to vernal equinox.

The impact solutions of (144898) 2004 VD17 in Tab. 6 were computed using 891 optical observations (of which 1 was rejected as outlier) from 2002/02/16.462 to 2006/04/22.871. These observations were available up to date of MACE 2006.

Tab. 6 shows that the infuence of σ and weighting of observations has a small influence for impact solutions for (144898) 2004 VD17.

Date	Dist. [RE]	Impact probability	Source
2102/05/04.894	0.51	6.66E-4	CLOMON2
2102/05/04.894	0.44	6.7 E-4	SENTRY
-	-		IW, 1 σ
2102/05/04.894	0.52	6.71E-4	IW, 1.5 σ
2102/05/04.894	0.52	6.22E-4	IW, 1.5 σ ,w=1
2102/05/04.894	0.52	6.71E-4	IW, 2 σ
2102/05/04.894	0.52	6.71E-4	IW, 2 σ, w=1
2102/05/04.894	0.52	6.66E-4	IW, 3 σ
2102/05/04.894	0.52	6.22E-4	IW, 3 σ, w=1
2102/05/04.894	0.52	6.30E-4	IW, 3 σ, no scal.
2102/05/04.894	0.52	9.37E-4	IW, 3 σ, fn
2102/05/04.894	0.52	6.71E-4	IW, 4 σ
2102/05/04.894	0.52	6.71E-4	IW, 5 σ
2102/05/04.894	0.52	6.71E-4	IW, 6 σ
2104/05/04.373	0.58	3.26E-7	CLOMON2
-	-		SENTRY
-	-		IW, 1 σ
2104/05/04.372	1.05	3.29E-7	IW, 1.5 σ
2104/05/04.377	1.15	3.14E-7	IW, 1.5 σ ,w=1
2104/05/04.374	0.54	3.29E-7	IW, 2 σ
2104/05/04.374	0.54	3.29E-7	IW, 2 σ, w=1
2104/05/04.373	0.74	3.29E-7	IW, 3 σ
2104/05/04.374	0.52	3.10E-7	IW, 3 σ, w=1
2104/05/04.374	0.55	3.15E-7	IW, 3 σ, no scal.
2104/05/04.375	0.58	4.80E-7	IW, 3 σ, fn
2104/05/04.374	0.52	3.31E-7	IW, 4 σ
2104/05/04.376	0.88	3.38E-7	IW, 5 σ
2104/05/04.374	0.54	3.36E-7	IW, 6 σ
2105/05/04.655	0.41	3.84E-8	IW, 3 σ, no scal.
2109/05/04.637	0.62	9.72E-9	IW, 1.5 σ, w=1

Table 6. (144898) 2004 VD17: Inluence of different *sigma_LOV* (σ) and weighting for impact solutions.

Mainly it have an effect on value of impact probability. Similar the problem of scaling of LOV (Milani et al., 2002) is neglecting in this case. Otherwise everywhere weighing is as CLOMON2, further settings are: multiple solution, use scaling (fn denotes impact solution without scaling), LOV with the largest eigenvalue; $w=1$ denotes without weighing of observations. On the MPML (Minor Planet Mailing List) forum the problem was connected with 4 first observations of (144898) 2004 VD17 recovered from 2002 year. It was appear that adding these observations does not affect on impact solutions considerably. In Tab. 6 *fn* denotes impact solutions without first four observations from 2002.

3.2 (144898) 2004 VD17: Approaching asteroids

As for (99942) Apophis, to compute exactly impact solutions for (144898) 2004 VD17 it is necessary to include gravitational perturbations of approaching asteroids. Using software Solex90 we have computed all close approaches of about 140,000 numbering asteroids known in Sept. 2006 with (144898) 2004 VD17 till 2110 year. We have found 5 asteroids with several close approaches to (144898) 2004 VD17 : (3) Juno, (6) Hebe, (7) Iris, (18) Melpomene and (51) Nemausa. These selected asteroids with the 4 massive ones were included to equations of motion of (144898) 2004 VD17 .

The computations of infuence of gravitational perturbations of these asteroids for the motion of (144898) 2004 VD17 were performed using software OrbFit 3.3.1. The masses of asteroids were taken from Michalak (2001) and from Solex90 as computed by A. Vitagliano (2006). The computations were based on 902 optical observations (of which 3 are rejected as outliers) from 2002/02/16.462 to 2006/04/29.090.

Date	Dist. [RE]	Impact probability	Source
2102/05/04.894	0.51	5.58 E-04	CLOMON2
2102/05/04.894	0.52	5.53 E-04	IW-a
2102/05/04.894	0.52	5.61 E-04	IW-d
2102/05/04.894	0.52	5.54 E-04	IW-b
2102/05/04.893	0.53	6.37 E-04	IW-bnrel
2102/05/04.894	0.52	5.59 E-04	IW-c
2102/05/04.894	0.52	5.53 E-04	IW-f
2102/05/04.890	0.44	7.35 E-04	SENTRY
2103/05/05.130	0.96	1.48 E-08	CLOMON2
2103/05/05.132	0.74	1.52 E-08	IW-b
2104/05/04.376	0.91	2.77 E-07	CLOMON2
2104/05/04.374	0.53	2.77 E-07	IW-a
2104/05/04.374	0.53	2.76 E-07	IW-d
2104/05/04.372	1.08	2.68 E-07	IW-b
2104/05/04.373	0.61	3.08 E-07	IW-bnrel
2104/05/04.376	0.87	2.78 E-07	IW-c
2104/05/04.377	1.09	2.74 E-07	IW-f
2109/05/04.515	0.84	7.60 E-09	IW-f

Table 7. (144898) 2004 VD17: Influence of approaching asteroids on impact solutions.

The results are in Tab. 7 where:

IW-a: solutions with 3. perturbing asteroids: Ceres, Pallas and Vesta

IW-b: 4 perturbing asteroids: Ceres, Pallas, Vesta, Hygiea

IW-bnrel: as *IW-b* without relativistic effects included

IW-c: 5 close approaching asteroids to (144898) 2004 VD17 : Juno, Hebe, Iris, Melpomene and Nemausa

IW-d: no perturbing asteroids

IW-f: all 9 perturbing asteroids: Ceres, Pallas, Juno, Vesta, Hebe, Iris, Hygiea, Melpomene and Nemausa

All results in Tab. 7 are computed with DE405 ephemeris and using relativistic effects (without case *IW-bnrel*). We can see that the impact solutions for asteroid (144898) 2004 VD17 does not differ so much using different number of perturbing asteroids as in the case of (99942) Apophis.

Fig. 4 shows the changes of differences in mean anomaly between asteroid (144898) 2004 VD17 on nominal orbits for different cases. In Fig. 4 (a) there are differences in mean anomaly between (144898) 2004 VD17 with and no relativistic effects included. Fig. 4 (b) presents differences in mean anomaly of (144898) 2004 VD17 between orbits without perturbing asteroids (solution *IW-d*) and with ones: 1 - solution *IW-a* (Ceres, Pallas and Vesta included), 2 - *IW-b* (Ceres, Pallas, Vesta, Hygiea), 3 - *IW-c* (Juno, Hebe, Iris, Melpomene, Nemausa), 4 - *IW-f* (Ceres, Pallas, Juno, Vesta, Hebe, Iris, Hygiea, Melpomene, Nemausa). The curves 1, 2 and 4 on the Fig. 4 are very similar, then the most perturbing effect comes from Ceres, Pallas and Vesta.

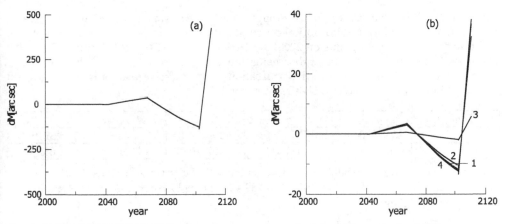

Fig. 4. (144898) 2004 VD17. Differences in mean anomaly between nominal orbit from different solutions: (a) – 4 perturbing asteroids: relativistic/non relativistic effects included; (b) – different number of perturbing asteroids: 1 - Ceres, Pallas and Vesta; 2 – Ceres, Pallas, Vesta and Hygiea; 3 – Juno, Hebe, Iris, Melpomene and Nemausa; 4 – Ceres, Pallas, Juno, Vesta, Hebe, Iris, Hygiea, Melpomene and Nemausa (see text) .

As in the case of (99942) Apophis the greatest infuence for motion (144898) 2004 VD17 have relativistic effects, about 10 times greater than the perturbing effects of additional massive asteroids. Even so we must use perturbing massive asteroids for computed precise impact solutions as Tab. 7 states. The rapidly changes in differences in mean anomaly in Fig. 4 are connected with the close approaches of (144898) 2004 VD17 with the Earth in the years: 2041 (0.01 AU), 2067 (0.03 AU) and 2102 (0.03 AU) for the nominal orbits. Hence chaoticity of the motion of the asteroid appears similar to this of (99942) Apophis but in the case of (144898) 2004 VD17 motion is less influenced.

3.3 (144898) 2004 VD17: The JPL Ephemerides

As in the case of (99942) Apophis using JPL Ephemerides DE405 and DE406 does not affect on the computed impact solutions in this short about 100 years time span.

4. Some new impact solution for (99942) Apophis and (144898) 2004 VD17

The new versions of the NEODyS-2: http://newton.dm.unipi.it/neodys/index.php?pc=0

and the new version of the OrbFit software: http://adams.dm.unipi.it/~orbmaint/orbfit/ were appeared.

Both they are based on the new error model (Chesley, Baer and Monet, 2010). The orbits are computed using star catalog debiasing and an error model with assumed astrometric errors RMS deduced from the tests of the paper cited above.

Also additional observations of (99942) Apophis and (144898) 2004 VD17 were added.

4.1 (99942) Apophis

Table 8 lists orbital elements of (99942) Apophis performed using all observations available to 1st Oct., 2011. There were 1481 optical observations (of which 8 were rejected as outliers), and also seven radar data points on 2005/01/27, 2005/01/29, 2005/01/31 and 2005/08/07. The orbit was computed by the author using the OrbFit software v. 4.2. The JPL NASA Ephemerides DE405 and additional perturbations from massive asteroids: (1) Ceres, (2) Pallas, (3) Juno, (4) Vesta and (10) Hygiea were used.

$M[deg]$	$a[AU]$	e	$\omega_{2000}[deg]$	$\Omega_{2000}[deg]$	$i_{2000}[deg]$
287.582224	0.9223003	0.19107616	126.42451	204.430372	3.331956
4.82E-05	1.13E-08	5.11E-08	7.57E-05	7.61E-05	1.63E-06

Table 8. (99942) Apophis: orbital elements together with theirs 1-σ variations. 1488 observations from 2555 days (2004/03/15.10789 – 2011/03/14.12528), rms=0.389". Nominal orbit: epoch 2011 Aug. 27.0.

Actually, both the Yarkovsky/YORP effect, which are part of a set of other astrodynamical effects that were taken summary only into account to prepare the present analysis, but that seems to be of significant influence in the orbital evolution of such objects. The preliminary results are in Table 9. The Yarkovsky and YORP (Yarkovsky-O' Keefe-Radzievskii-Paddack) effects are thermal radiation forces and torques that cause a drift of semimajor axes (computed value of **da/dt** in present work) of small asteroids and meteoroids and a change their spin vectors (obliquities). Because the Yarkovsky force depends on the obliquity, we can expect a complicated interplay between the Yarkovsky and YORP effects . Therefore it is difficult to estimate the influence of the Yarkovsky and YORP effects on the motion of asteroids separately. The result of the Yarkovsky effect is removal of small asteroids from the main belt to chaotic mean motion and secular apsidal or nodal resonance zones. Then they can be gradually transported to Earth-crossing orbits. Therefore the Yarkovsky and YORP effects are now considered in relation to objects crossing the Earth orbit, particularly they are important in the motion of potentially dangerous asteroids for the Earth.

The NEODyS presents only impact solutions based on 1399 optical observations (of which 5 are rejected as outliers) from 2004/03/15.127 to 2008/01/09.666 and also on seven radar data points on 2005/01/27, 2005/01/29, 2005/01/31, 2005/08/07 and 2006/05/06. The NEODyS lists possible impact in 2036, 2056, 2068 – two solutions, 2076 and in 2103. Their solutions are based on Monte Carlo method, including a probabilistic model of the Yarkovsky effect. In this way impact solutions are model dependent. Without any non-gravitational perturbation model they found possible impact in 2068/04/12.632 with the probability of about $3.81 \ 10^{-6}$. Similar impact solutions are from the JPL NASA.

Using all 1490 observations of Apophis and the OrbFit software I computed value of the semimajor axis drift of Apophis equal to **da/dt**=+180 10^{-4} AU/Myr connected with the Yarkovsky/YORP effects and got following impact solutions as are presented in Table 9. Additional perturbations from (1) Ceres, (2) Pallas and (4) Vesta are included.

Date	Dist. [RE]	Impact probability
2043/04/13.901	2.01	2.08 E-07
2064/04/12.824	1.52	1.44 E-06

Table 9. (99942) Apophis. Impact solutions with the Earth using semimajor axis drif, **da/dt**= =+180 10^{-4} AU/Myr, computed by the author. Similar value of **da/dt**= (235+/-50) 10^{-4} AU/Myr has computed Grzegorz Sitarski (private information).

To compute impact solutions of Apophis we must know exact uncertainty from the Yarkovsky effects and physical parameter uncertainties of Apophis together with the astrometric biases and radiation pressure (Giorgini et al. 2007).

4.2 (144898) 2004 VD17

Table 10 lists orbital elements of (144898) 2004 VD17 performed using all observations available up to 1st Oct., 2011. There were 981 optical observations of which 4 were rejected as outliers. The orbit was computed by the author using the OrbFit software v. 4.2. The JPL NASA Ephemerides DE405 and additional perturbations from massive asteroids: (1) Ceres, (2) Pallas, (3) Juno, (4) Vesta and (10) Hygiea were used.

$M[deg]$	$a[AU]$	e	$\omega_{2000}[deg]$	$\Omega_{2000}[deg]$	$i_{2000}[deg]$
271.359237	1.5081219	0.58860168	90.762748	224.187119	4.223233
9.96E-05	7.36E-08	4.86E-08	3.52E-05	7.93E-06	7.93E-06

Table 10. (144898) 2004 VD17: orbital elements together with theirs 1-σ variations. 981 observations from 2513 days (2002/02/16.46212 – 2009/01/03.40647), *rms*=0.402". Nominal orbit: epoch 2011 Aug. 27.0.

No impact solutions were found for (144898) VD17 in the next 100 year in the future. Similar solution, no possible impact, were detected by the JPL NASA and by the NEODyS.

5. Conclusion

To compute precisely the impact solutions of (99942) Apophis and (144898) 2004 VD17 it is necessary to include small effects like relativistic effects, close approaching asteroids, the Yarkovsky/YORP effect. The use of the software OrbFit is helpful in computing exact possible impacts of asteroids with the Earth. Thanks for the OrbFit Consortium. Also the free software Solex was useful in this work.

6. Acknowledgment

Thank you to the researches from OrbFit Consortium working in four research laboratories: http://adams.dm.unipi.it/~orbmaint/orbfit/OrbFit/doc/help.html#authors for theirs free software and source code.

Thank you very much for Andrea Milani and Geny Sansaturio for discussions during MACE 2006 and for the help through e-mails.

I also thank Aldo Vitagliano for his discussion about Solex during Mace 2006.

I would like to thank Fabrizio Bernardi for his help in the OrbFit v.4.2.

I would like to thank Grzegorz Sitarski from the Space Research Center, of the Polish Academy of Sciences in Warsaw for his impact solutions of (99942) Apophis and (144898) 2004 VD17 and profound and fruitful help.

I am also grateful for the suggestions given by Leonard Kornos.

7. References

Chesley, S. R., Baer, J., Monet, D. G. (2010). *Icarus*, 210, 158
Giorgini, J. D., Benner, L. A. M., Ostro, S. J., Nolan, M. C., Busch, M. W. (2008). *Icarus*, 193, 1
Michalak, G. (2001). *Astron. Astrophys.* 374, 703
Milani, A., Chesley, S. R., Chodas, P. W., Valsecchi, G. B. (2002). In *Asteroids III*, ed.: W.F. Bottke Jr., A. Cellino, P. Paolicchi, and R. P. Binzel (eds), Univ. of Arizona Press, Tucson, 55
Milani, A., Chesley, S.R., Sansaturio, M. E., Tommei, G., Valsecchi, G.B. (2005a). *Icarus* 173, 362
Milani, A., Sansaturio, M. E., Tommei, G., Arratia, O., Chesley, S. R. (2005b). *Astron. Astrophys.* 431, 729
Sitarski, G. (2002). *Acta Astron.* 52, 471
Vitagliano, A. (1997). *Cel. Mech. Dyn. Astron.* 66, 293
Vitagliano, A. and Stoss, R. (2006). *Astron. Astrophys.* 455, L29
Wlodarczyk, I. (2001). *Acta Astron.* 51, 357
Yeomans, D.K., Ostro, S.J., Chodas, P.W. (1987). *Astron. J.* 94, 189

5

Secular Evolution of Satellites by Tidal Effect

Alexandre C. M. Correia
University of Aveiro
Portugal

1. Introduction

Both the Earth's Moon and Pluto's moon, Charon, have an important fraction of the mass of their systems, and therefore they could be classified as double-planets rather than as satellites. The proto-planetary disk is unlikely to produce such systems, and their origin seems to be due to a catastrophic impact of the initial planet with a body of comparable dimensions (e.g. Canup, 2005; Canup & Asphaug, 2001). On the other hand, Neptune's moon, Triton, and the Martian moon, Phobos, are spiraling down into the planet, clearly indicating that the present orbits are not primordial, and may have undergone a long evolving process from a previous capture (e.g. Goldreich et al., 1989; Mignard, 1981).

The present orbits of all these satellites are almost circular, and their spins appear to be synchronous with the orbital mean motion, as well as being locked in Cassini states (e.g. Colombo, 1966; Peale, 1969). This also applies to the Galilean satellites of Jupiter, which are likely to have originated from Jupiter's accretion disk and additionally show orbital mean motion resonances (e.g. Yoder, 1979). All these features seem to be due to tidal evolution, which arises from differential and inelastic deformation of the planet by a perturbing body.

Previous long-term studies on the orbital evolution of satellites have assumed that their rotation is synchronously locked, and therefore limits the tidal evolution to the orbits (e.g. McCord, 1966). However, these two kinds of evolution cannot be dissociated because the total angular momentum must be conserved. Additionally, it has been assumed that the spin axis is locked in a Cassini state with very low obliquity. Although these assumptions are correct for the presently known situations, they were not necessarily true throughout the evolution.

In this article we model the orbital evolution of a satellite from its origin or capture until the preset day, including spin evolution for both planet and satellite, and we also regard its future evolution. We provide a simple averaged model adapted for fast computational simulations, as required for long-term studies, following Correia (2009). However, we present an improvement with respect to previous work, here we do not average the equations of motion over the argument of the periastron, as in Correia et al. (2011). Therefore, this model is more complete, and allows the eccentricity of the satellite to show secular variations due to the gravitational perturbations of the star on its orbit around the planet. We then apply this model to the Triton-Neptune system. The results do not differ much from those in Correia (2009) for the final stages of the orbital evolution, but can show some significant differences during the initial stages. In the last section we discuss the results obtained.

2. The model

We consider a hierarchical system composed of a star, a planet and a satellite, with masses $M \gg m_0 \gg m_1$, respectively. Both planet and satellite are considered oblate ellipsoids with gravity field coefficients given by J_{2_0} and J_{2_1}, rotating about the axis of maximal inertia along the directions $\hat{s}_0$ and $\hat{s}_1$, with rotation rates ω_0 and ω_1, respectively. The potential energy U of the system is then given by (e.g. Smart, 1953):

$$
U = -G \frac{M m_0}{r_0} \left(1 - \sum_{i=0,1} J_{2_i} \frac{m_i}{m_0} \left(\frac{R_i}{r_0} \right)^2 P_2(\hat{r}_0 \cdot \hat{s}_i) \right)
$$

$$
- G \frac{m_0 m_1}{r_1} \left(1 - \sum_{i=0,1} J_{2_i} \left(\frac{R_i}{r_1} \right)^2 P_2(\hat{r}_1 \cdot \hat{s}_i) \right)
$$

$$
- G \frac{M m_1}{r_0} \left(\frac{r_1}{r_0} \right)^2 P_2(\hat{r}_0 \cdot \hat{r}_1) , \tag{1}
$$

where terms in $(R_i/r_j)^3$ have been neglected $(i, j = 0, 1)$. G is the gravitational constant, R_i the radius of the planet or the satellite, r_i the distance between the planet and the star or the satellite, and $P_2(x) = (3x^2 - 1)/2$ the Legendre polynomial of degree two.

Neglecting tidal interactions with the star, the tidal potential is written (e.g. Kaula, 1964):

$$
U_T = - \frac{G}{r_1^3} \sum_{i=0,1} k_{2_i} m_{(1-i)}^2 \frac{R_i^5}{r_i'^3} P_2(\hat{r}_1 \cdot \hat{r}_i') , \tag{2}
$$

where k_{2_i} is the potential Love number for the planet or the satellite, and r_i' the position of the interacting body at a time delayed of Δt_i. For simplicity, we will adopt a model with constant Δt_i, which can be made linear (e.g. Mignard, 1979; Néron de Surgy & Laskar, 1997):

$$
r_i' \simeq r_1 + \Delta t_i (\omega_i s_i \times r_1 - \dot{r}_1) . \tag{3}
$$

The complete evolution of the system can be tracked by the evolution of the rotational angular momentums, $\mathbf{H}_i \simeq C_i \omega_i \hat{s}_i$, the orbital angular momentums, $\mathbf{L}_i \simeq m_i n_i a_i^2 (1 - e_i^2)^{1/2} \hat{k}_i$, and the Laplace-Runge-Lenz vector, which points along the major axis in the direction of periapsis with magnitude e_1:

$$
\mathbf{e}_1 = \frac{\dot{\mathbf{r}}_1 \times \mathbf{L}_1}{G M m_1} - \frac{\mathbf{r}_1}{r_1} . \tag{4}
$$

n_i is the mean motion, a_i the semi-major axis, e_i the eccentricity, and C_i the principal moment of inertia. The contributions to the orbits are easily computed from the above potentials as

$$
\dot{\mathbf{L}}_0 = -\mathbf{r}_0 \times \mathbf{F}_0 , \quad \dot{\mathbf{L}}_1 = -\mathbf{r}_1 \times \mathbf{F}_1 , \tag{5}
$$

$$
\dot{\mathbf{e}}_1 = \frac{1}{G M m_1} \left(\mathbf{F}_1 \times \frac{\mathbf{L}_1}{m_1} + \dot{\mathbf{r}}_1 \times \dot{\mathbf{L}}_1 \right) , \tag{6}
$$

where $\mathbf{F}_i = -\nabla_{\mathbf{r}_i} U'$, with $U' = U + U_T + G M m_0/r_0 + G m_0 m_1/r_1$.

Since the total angular momentum is conserved, the contributions to the spin of the planet and satellite can easily be computed from the orbital contributions:

$$\dot{\mathbf{H}}_0 + \dot{\mathbf{H}}_1 + \dot{\mathbf{L}}_0 + \dot{\mathbf{L}}_1 = 0 . \tag{7}$$

Because we are only interested in the secular evolution of the system, we further average the equations of motion over the mean anomalies of both orbits. The resulting equations for the conservative motion are (Boué & Laskar, 2006; Farago & Laskar, 2010):

$$\dot{\mathbf{L}}_0 = -\gamma(1-e_1^2)\cos I\,\hat{\mathbf{k}}_1 \times \hat{\mathbf{k}}_0 + 5\gamma(\mathbf{e}_1 \cdot \hat{\mathbf{k}}_0)\,\mathbf{e}_1 \times \hat{\mathbf{k}}_0 - \sum_i \alpha_i \cos\varepsilon_i\,\hat{\mathbf{s}}_i \times \hat{\mathbf{k}}_0 , \tag{8}$$

$$\dot{\mathbf{L}}_1 = -\gamma(1-e_1^2)\cos I\,\hat{\mathbf{k}}_0 \times \hat{\mathbf{k}}_1 + 5\gamma(\mathbf{e}_1 \cdot \hat{\mathbf{k}}_0)\,\hat{\mathbf{k}}_0 \times \mathbf{e}_1 - \sum_i \beta_i \cos\theta_i\,\hat{\mathbf{s}}_i \times \hat{\mathbf{k}}_1 , \tag{9}$$

$$\dot{\mathbf{e}}_1 = -\frac{\gamma(1-e_1^2)}{\|\mathbf{L}_1\|}\left[\cos I\,\hat{\mathbf{k}}_0 \times \mathbf{e}_1 - 2\,\hat{\mathbf{k}}_1 \times \mathbf{e}_1 - 5(\mathbf{e}_1 \cdot \hat{\mathbf{k}}_0)\,\hat{\mathbf{k}}_0 \times \hat{\mathbf{k}}_1\right]$$

$$-\sum_i \frac{\beta_i}{\|\mathbf{L}_1\|}\left[\cos\theta_i\,\hat{\mathbf{s}}_i \times \mathbf{e}_1 + \frac{1}{2}(1 - 5\cos^2\theta_i)\,\hat{\mathbf{k}}_1 \times \mathbf{e}_1\right] , \tag{10}$$

and

$$\dot{\mathbf{H}}_i = -\alpha_i \cos\varepsilon_i\,\hat{\mathbf{k}}_0 \times \hat{\mathbf{s}}_i - \beta_i \cos\theta_i\,\hat{\mathbf{k}}_1 \times \hat{\mathbf{s}}_i , \tag{11}$$

where

$$\alpha_i = \frac{3GMm_i J_{2_i} R_i^2}{2a_0^3(1-e_0^2)^{3/2}} , \tag{12}$$

$$\beta_i = \frac{3Gm_0 m_1 J_{2_i} R_i^2}{2a_1^3(1-e_1^2)^{3/2}} , \tag{13}$$

$$\gamma = \frac{3GMm_1 a_1^2}{4a_0^3(1-e_0^2)^{3/2}} , \tag{14}$$

and

$$\cos\varepsilon_i = \hat{\mathbf{s}}_i \cdot \hat{\mathbf{k}}_0 , \quad \cos\theta_i = \hat{\mathbf{s}}_i \cdot \hat{\mathbf{k}}_1 , \quad \cos I = \hat{\mathbf{k}}_0 \cdot \hat{\mathbf{k}}_1 , \tag{15}$$

are the direction cosines of the spins and orbits: ε_i is the obliquity to the orbital plane of the planet, θ_i the obliquity to the orbital plane of the satellite, and I the inclination between orbital planes.

For the dissipative tidal effects, we obtain (Correia et al., 2011):

$$\dot{\mathbf{L}}_0 = 0 , \quad \dot{\mathbf{L}}_1 = -\dot{\mathbf{H}}_0 - \dot{\mathbf{H}}_1 , \tag{16}$$

$$\dot{\mathbf{e}}_1 = \sum_i \frac{15}{2}k_{2_i}n_1\left(\frac{m_j}{m_i}\right)\left(\frac{R_i}{a_1}\right)^5 f_4(e_1)\,\hat{\mathbf{k}}_1 \times \mathbf{e}_1$$

$$-\sum_i \frac{K_i}{\beta_1 a_1^2}\left[f_4(e_1)\frac{\omega_i}{2n_1}(\mathbf{e}_1 \cdot \hat{\mathbf{s}}_i)\,\hat{\mathbf{k}}_1 - \left(\frac{11}{2}f_4(e_1)\cos\theta_i\frac{\omega_i}{n_1} - 9f_5(e_1)\right)\mathbf{e}_1\right] , \tag{17}$$

and

$$\dot{\mathbf{H}}_i = K_i\, n_1 \left[f_4(e_1)\sqrt{1 - e_1^2}\,\frac{\omega_i}{2n_1}\,(\hat{\mathbf{s}}_i - \cos\theta_i\,\hat{\mathbf{k}}_1)\right.$$

$$\left. -f_1(e_1)\frac{\omega_i}{n_1}\hat{\mathbf{s}}_i + f_2(e_1)\hat{\mathbf{k}}_1 + \frac{(\mathbf{e}_1 \cdot \hat{\mathbf{s}}_i)(6 + e_1^2)}{4(1 - e_1^2)^{9/2}}\frac{\omega_i}{n_1}\mathbf{e}_1 \right], \tag{18}$$

where,

$$K_i = \Delta t_i\,\frac{3k_{2_i}\, Gm_{(1-i)}^2\, R_i^5}{a_1^6}, \tag{19}$$

and

$$f_1(e) = \frac{1 + 3e^2 + \frac{3}{8}e^4}{(1 - e^2)^{9/2}}, \tag{20}$$

$$f_2(e) = \frac{1 + \frac{15}{2}e^2 + \frac{45}{8}e^4 + \frac{5}{16}e^6}{(1 - e^2)^6}, \tag{21}$$

$$f_3(e) = \frac{1 + \frac{31}{2}e^2 + \frac{255}{8}e^4 + \frac{185}{16}e^6 + \frac{25}{64}e^8}{(1 - e^2)^{15/2}}, \tag{22}$$

$$f_4(e) = \frac{1 + \frac{3}{2}e^2 + \frac{1}{8}e^4}{(1 - e^2)^5}, \tag{23}$$

$$f_5(e) = \frac{1 + \frac{15}{4}e^2 + \frac{15}{8}e^4 + \frac{5}{64}e^6}{(1 - e^2)^{13/2}}. \tag{24}$$

The first term in expression (17) corresponds to a permanent tidal deformation, while the second term corresponds to the dissipative contribution. The precession rate of $\mathbf{e}_1$ about $\hat{\mathbf{k}}_1$ is usually much faster than the evolution time-scale for the dissipative tidal effects. As a consequence, when the eccentricity is constant over a precession cycle, we can average expression (18) over the argument of the periapsis and get (Correia, 2009):

$$\dot{\mathbf{H}}_i = -K_i\, n_1 \left(f_1(e_1)\frac{\hat{\mathbf{s}}_i + \cos\theta_i\,\hat{\mathbf{k}}_1}{2}\frac{\omega_i}{n_1} - f_2(e_1)\hat{\mathbf{k}}_1 \right). \tag{25}$$

3. Secular evolution

In the previous section we presented the equations that rule the tidal evolution of a satellite in terms of angular momenta and orbital energy. However, the spin and orbital quantities are better represented by the rotation angles and elliptical elements. The direction cosines (Eq.15) are obtained from the angular momenta vectors, since $\hat{\mathbf{s}}_i = \mathbf{H}_i/||\mathbf{H}_i||$ and $\hat{\mathbf{k}}_i = \mathbf{L}_i/||\mathbf{L}_i||$, as well as the rotation rate $\omega_i = \mathbf{H}_i \cdot \hat{\mathbf{s}}_i/C_i$. The eccentricity and the semi-major axis can be obtained from $e_1 = ||\mathbf{e}_1||$, and $a_1 = ||\mathbf{L}_1||^2/(GMm_1^2(1 - e_1^2))$, respectively.

3.1 Spin evolution

The variation in the satellite's rotation rate can be obtained from $\dot{\omega}_i = \dot{\mathbf{H}}_i \cdot \hat{\mathbf{s}}_i / C_i$ (Eq. 25), giving (Correia & Laskar, 2009):

$$\dot{\omega}_1 = -\frac{K_1 n_1}{C_1} \left(f_1(e_1) \frac{1 + \cos^2 \theta_1}{2} \frac{\omega_1}{n_1} - f_2(e_1) \cos \theta_1 \right) . \tag{26}$$

For a given obliquity and eccentricity, the equilibrium rotation rate, obtained when $\dot{\omega}_1 = 0$, is attained for:

$$\frac{\omega_1}{n_1} = \frac{f_2(e_1)}{f_1(e_1)} \frac{2 \cos \theta_1}{1 + \cos^2 \theta_1} , \tag{27}$$

The obliquity variations can be obtained from equation (15):

$$\frac{d \cos \theta_i}{dt} = \frac{\dot{\mathbf{H}}_i \cdot (\hat{\mathbf{k}}_1 - \cos \theta_i \hat{\mathbf{s}}_i)}{||\mathbf{H}_i||} + \frac{\dot{\mathbf{L}}_1 \cdot (\hat{\mathbf{s}}_i - \cos \theta_i \hat{\mathbf{k}}_1)}{||\mathbf{L}_1||} . \tag{28}$$

For the conservative motion (Eqs. 9, 11), stable configurations for the spin can be found whenever the vectors $(\hat{\mathbf{s}}_1, \hat{\mathbf{k}}_1, \hat{\mathbf{k}}_0)$ or $(\hat{\mathbf{s}}_1, \hat{\mathbf{k}}_1, \hat{\mathbf{s}}_0)$ are coplanar and precess at the same rate g (e.g. Colombo, 1966; Correia et al., 2011; Peale, 1969). The first situation occurs if $\gamma \gg \beta_0$ (outer satellite) and the second situation when $\gamma \ll \beta_0$ (inner satellite). The equilibrium obliquities can be found from a single relationship (e.g. Ward & Hamilton, 2004):

$$\lambda_1 \cos \theta_1 \sin \theta_1 + \sin(\theta_1 - I_0) = 0 , \tag{29}$$

where $\lambda_1 = \beta_1 / (C_1 \omega_1 g)$ is a dimensionless parameter and I_0 is the inclination of the orbit of the satellite with respect to the Laplacian plane ($I_0 \simeq I$ and $g \simeq \gamma \cos I / ||\mathbf{L}_1||$ for an outer satellite, and $I_0 \simeq \theta_0$ and $g \simeq \beta_0 \cos \theta_0 / ||\mathbf{L}_1||$ for an inner satellite) (e.g. Laplace, 1799; Mignard, 1981; Tremaine et al., 2009). The above equation has two or four real roots for θ_1, which are known by *Cassini states*. In general, for satellites we have $I_0 \sim 0$, and these solutions are approximately given by:

$$\tan^{-1} \left(\frac{\sin I_0}{\cos I_0 \pm \lambda_1} \right) , \quad \pm \cos^{-1} \left(-\frac{\cos I_0}{\lambda_1} \right) . \tag{30}$$

For a generic value of I_0, when $\lambda_1 \ll 1$, which is often the case of an outer satellite, the first expression gives the only two real roots of equation (29), one for $\theta_1 \simeq I_0$ and another for $\theta_1 \simeq \pi - I_0$. On the other hand, when $\lambda_1 \gg 1$, which is the case of inner satellites, we will have four real roots approximately given by expressions (30).

In turn, the dissipative obliquity variations are computed by substituting equation (25) in (28) with $||\mathbf{H}_1|| \ll ||\mathbf{L}_1||$, giving:

$$\dot{\theta}_1 \simeq \frac{K_1 n_1}{C_1 \omega_1} \sin \theta_1 \left(f_1(e_1) \cos \theta_1 \frac{\omega_1}{2 n_1} - f_2(e_1) \right) . \tag{31}$$

Because of the factor n_1 / ω_1 in the magnitude of the obliquity variations, for an initial fast rotating satellite, the time-scale for the obliquity evolution will be longer than the time-scale for the rotation rate evolution (Eq.26). As a consequence, it is to be expected that the rotation rate reaches its equilibrium value (Eq.27) earlier than the obliquity. Thus, replacing equation (27) in (31), we have:

$$\dot{\theta}_1 \simeq -\frac{K_1 n_1}{C_1 \omega_1} f_2(e_1) \frac{\sin \theta_1}{1 + \cos^2 \theta_1} . \tag{32}$$

We then conclude that the obliquity can only decrease by tidal effect, since $\dot{\theta}_1 \leq 0$, and the final obliquity tends to be captured in low obliquity Cassini states.

3.2 Orbital evolution

The variations in the eccentricity are easily obtained from the Laplace-Runge-Lenz vector (Eq.17):

$$\dot{e}_1 = \frac{\dot{\mathbf{e}}_1 \cdot \mathbf{e}_1}{e_1} = \sum_i \frac{9K_i}{m_1 a_1^2} \left(\frac{11}{18} f_4(e_1) \cos \theta_i \frac{\omega_i}{n_1} - f_5(e_1) \right) e_1 , \tag{33}$$

while the semi-major axis variations are obtained from the eccentricity and the norm of the orbital angular momentum:

$$\frac{\dot{a}_1}{a_1} = \frac{2\dot{e}_1 e_1}{(1 - e_1^2)} + \frac{2\dot{\mathbf{L}}_1 \cdot \mathbf{L}_1}{||\mathbf{L}_1||^2} = \sum_i \frac{2K_i}{m_1 a_1^2} \left(f_2(e_1) \cos \theta_i \frac{\omega_i}{n_1} - f_3(e_1) \right) . \tag{34}$$

For gaseous planets and rocky satellites we usually have $k_{2_0} \Delta t_0 \ll k_{2_1} \Delta t_1$, and we can retain only terms in K_1.

The ratio between orbital and spin evolution time-scales is roughly given by $C_1 / (m_1 a_1^2) \ll 1$, meaning that the spin achieves an equilibrium position ($\dot{\mathbf{H}}_1 = 0$) much faster than the orbit. Replacing the equilibrium rotation rate (Eq. 27) with $\theta_1 = 0$ (for simplicity) in equations (33) and (34), gives:

$$\dot{a}_1 = -\frac{7K_1}{m_1 a_1} f_6(e_1) e_1^2 , \tag{35}$$

$$\dot{e}_1 = -\frac{7K_1}{2m_1 a_1^2} f_6(e_1)(1 - e_1^2) e_1 , \tag{36}$$

where

$$f_6(e) = \frac{1 + \frac{45}{14}e^2 + 8e^4 + \frac{685}{224}e^6 + \frac{255}{448}e^8 + \frac{25}{1792}e^{10}}{(1 + 3e^2 + \frac{3}{8}e^4)(1 - e^2)^{15/2}} . \tag{37}$$

Thus, we always have $\dot{a}_1 \leq 0$ and $\dot{e}_1 \leq 0$, and the final eccentricity is zero. However, from this point onwards, the tidal effects on the planet cannot be neglected (Eq.34), and they govern the future evolution of the satellite's orbit. For $a_f < a_s$ or $\theta_0 \geq \pi/2$, where $a_s^3 = Gm_0 / (\omega_0 \cos \theta_0)^2$, the semi-major axis continues to decrease until the satellite crashes into the planet, while in the remaining situations it will increase.

4. Application to Triton-Neptune

Neptune's main satellite, Triton, presents unique features in the Solar System. It is the only moon-sized body in a retrograde inclined orbit and the images taken by the Voyager 2 spacecraft in 1989 revealed an extremely young surface with few impact craters (e.g. Cruikshank, 1995). This satellite should have remained molten until about 1 Gyr ago and

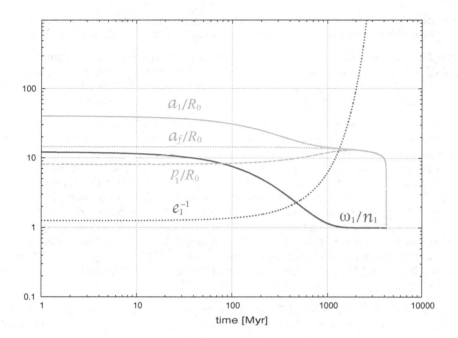

Fig. 1. Secular tidal evolution of the Triton-Neptune system. We plot the semi-major axis ratios a_1/R_0 and a_f/R_0, the periastron distance p_1/R_0, the inverse of the eccentricity e_1^{-1} and Triton's rotation rate ratio ω_1/n_1.

its interior is still warm and geologically active considering its distance from the Sun (Schenk & Zahnle, 2007). Its composition also presents some similarities with Pluto (Tsurutani et al., 1990).

These bizarre characteristics lead one to believe that Triton originally orbited the Sun, belonging to the family of Kuiper-belt objects. Most likely during the outward migration of Neptune, the orbits of the two bodies intercepted and capture occurred. This possibility is strongly supported by the fact that Triton's present orbit lies between a group of small inner prograde satellites and a number of exterior irregular satellites both prograde and retrograde. Nereid, with an orbital eccentricity around 0.75, is also believed to have been scattered from a regular satellite orbit (McKinnon, 1984).

How exactly the capture occurred is still unknown, but some mechanisms have been proposed: gas drag (McKinnon & Leith, 1995; Pollack et al., 1979), a collision with a pre-existing regular satellite of Neptune (Goldreich et al., 1989), or three-body interactions (Agnor & Hamilton, 2006; Vokrouhlický et al., 2008). All these scenarios require a very close passage to Neptune, and leave the planet in eccentric orbits that must be damped by tides to the present one. Tides are thus the only consensual mechanism acting on Triton's orbit. The tidal distortion of Triton after a few close passages around Neptune, and the consequent dissipation of tidal energy, can account for a substantial reduction in the semi-major axis of its orbit, quickly bringing the planet from an orbit outside Neptune's Hill sphere ($\sim 4700\, R_0$)

to a bounded orbit. Therefore, it cannot be ruled out that Triton was simply captured by tidal interactions with Neptune after a close encounter in an almost parabolic orbit (McCord, 1966).

Here, we simulate the tidal evolution of the Triton-Neptune system using the complete model described in Sect.2. Triton is started in a very elliptical orbit with $e_1 = 0.99$ and a semi-major axis of $a_1 = 40 R_0$, corresponding to a final equilibrium $a_f \simeq 14.4 R_0$, close to the present position of $14.33 R_0$. These specific values give a closest approach at a periapse of $8 R_0$.

For the radius of the bodies we use $R_0 = 24\,764$ km and $R_1 = 1\,353$ km (Thomas, 2000), while for the masses, the J_2 of Neptune and the remaining orbital and spin parameters we take the present values as determined by Jacobson (2009). For Triton we adopt $J_2 = 4.38 \times 10^{-4}$, the value measured for Europa (Anderson et al., 1998), and $C_{22} = 0$, since our model does not take into account spin-orbit resonances. This choice is justified because Triton's observed topography never varies beyond a kilometer (Thomas, 2000). In addition, Triton should have undergone frequent collisions either with other satellites of Neptune, or with external Kuiper-belt objects, and any capture in a spin-orbit resonance different from the synchronous one, may not last for a long time (Stern & McKinnon, 2000). For tidal dissipation we adopt the same parameters as previous studies, that is, $k_{2_0} = 0.407$ and $Q_0 = 9000$ (Zhang & Hamilton, 2008), and $k_{2_1} = 0.1$ and $Q_1 = 100$ (Chyba et al., 1989; Goldreich et al., 1989), where $Q_i^{-1} = \omega_i \Delta t_i$.

As for the orbit, the initial spin of Triton is unknown. We tested several possibilities, but tides acting on the spin always drive it in the same way: the rotation rate quickly evolves into the equilibrium value given by equation (27), while the obliquity is trapped in a Cassini state. In our standard simulation (Fig.1) we start Triton with a rotation period of 24 h. The semi-major axis and the eccentricity always decrease, as predicted by equations (35) and (36), and the quantity $a_f = a_1(1 - e_1^2)$ is preserved during the first stages of the evolution, with the reduction observed being caused by tides on Neptune. The eccentricity is very high during the first evolutionary stages, but it decreases rapidly as the satellite approaches its present orbit. Finally, the rotation of the satellite decreases as the satellite orbit shrinks into Neptune and ultimately stabilizes in the synchronous resonance, the presently observed configuration.

5. Conclusion

The numerical results presented here are very similar to those shown in Correia (2009) for $a_1 \leq 40 R_0$. For these values of the semi-major axis, Triton can still be considered as an "inner satellite", that is, the inclination with respect to the equatorial plane of Neptune, θ_0, is approximately constant. However, for higher values of the semi-major axis, we observe exchanges between the inclination and the eccentricity of Triton. Since the eccentricity is no longer constant, it is not possible to average over the argument of the periastron as in Correia (2009). Therefore, the results with the non-averaged model presented here will show some differences. In particular, the perturbation on Triton's orbit will cause the eccentricity to vary around the mean value, allowing the periapse to attain lower values. As a consequence, tidal effects will be stronger for close encounters with Neptune, and the migration of Triton may occur in much faster time-scales. In a future study we will analyze in detail the evolution of Triton's orbit for $a_1 > 40 R_0$.

Our study should also apply to the Moon, Charon and the satellites of Mars, although in this case we need to take into account the quadropole moment of inertia $C_{22} \neq 0$ (Correia, 2006).

6. References

Agnor, C. B. & Hamilton, D. P. (2006). Neptune's capture of its moon Triton in a binary-planet gravitational encounter, *Nature* 441: 192–194.

Anderson, J. D., Schubert, G., Jacobson, R. A., Lau, E. L., Moore, W. B. & Sjogren, W. L. (1998). Europa's Differentiated Internal Structure: Inferences from Four Galileo Encounters, *Science* 281: 2019–2022.

Boué, G. & Laskar, J. (2006). Precession of a planet with a satellite, *Icarus* 185: 312–330.

Canup, R. M. (2005). A Giant Impact Origin of Pluto-Charon, *Science* 307: 546–550.

Canup, R. M. & Asphaug, E. (2001). Origin of the Moon in a giant impact near the end of the Earth's formation, *Nature* 412: 708–712.

Chyba, C. F., Jankowski, D. G. & Nicholson, P. D. (1989). Tidal evolution in the Neptune-Triton system, *Astron. Astrophys.* 219: L23–L26.

Colombo, G. (1966). Cassini's second and third laws, *Astron. J.* 71: 891–896.

Correia, A. C. M. (2006). The core-mantle friction effect on the secular spin evolution of terrestrial planets, *Earth Planet. Sci. Lett.* 252: 398–412.

Correia, A. C. M. (2009). Secular Evolution of a Satellite by Tidal Effect: Application to Triton, *Astrophys. J.* 704: L1–L4.

Correia, A. C. M. & Laskar, J. (2009). Mercury's capture into the 3/2 spin-orbit resonance including the effect of core-mantle friction, *Icarus* 201: 1–11.

Correia, A. C. M., Laskar, J., Farago, F. & Boué, G. (2011). Tidal evolution of hierarchical and inclined systems, *Celestial Mechanics and Dynamical Astronomy* 111: 105–130.

Cruikshank, D. P. (1995). *Neptune and Triton*, University of Arizona Press.

Farago, F. & Laskar, J. (2010). High-inclination orbits in the secular quadrupolar three-body problem, *Mon. Not. R. Astron. Soc.* 401: 1189–1198.

Goldreich, P., Murray, N., Longaretti, P. Y. & Banfield, D. (1989). Neptune's story, *Science* 245: 500–504.

Jacobson, R. A. (2009). The Orbits of the Neptunian Satellites and the Orientation of the Pole of Neptune, *Astron. J.* 137: 4322–4329.

Kaula, W. M. (1964). Tidal dissipation by solid friction and the resulting orbital evolution, *Rev. Geophys.* 2: 661–685.

Laplace, P. S. (1799). *Traité de Mécanique céleste*, Paris: Gauthier-Villars.

McCord, T. B. (1966). Dynamical evolution of the Neptunian system, *Astron. J.* 71: 585–590.

McKinnon, W. B. (1984). On the origin of Triton and Pluto, *Nature* 311: 355–358.

McKinnon, W. B. & Leith, A. C. (1995). Gas drag and the orbital evolution of a captured Triton., *Icarus* 118: 392–413.

Mignard, F. (1979). The evolution of the lunar orbit revisited. I, *Moon and Planets* 20: 301–315.

Mignard, F. (1981). Evolution of the Martian satellites, *Mon. Not. R. Astron. Soc.* 194: 365–379.

Néron de Surgy, O. & Laskar, J. (1997). On the long term evolution of the spin of the Earth., *Astron. Astrophys.* 318: 975–989.

Peale, S. J. (1969). Generalized Cassini's Laws, *Astron. J.* 74: 483–489.

Pollack, J. B., Burns, J. A. & Tauber, M. E. (1979). Gas drag in primordial circumplanetary envelopes - A mechanism for satellite capture, *Icarus* 37: 587–611.

Schenk, P. M. & Zahnle, K. (2007). On the negligible surface age of Triton, *Icarus* 192: 135–149.

Smart, W. M. (1953). *Celestial Mechanics.*, London, New York, Longmans, Green.

Stern, S. A. & McKinnon, W. B. (2000). Triton's Surface Age and Impactor Population Revisited in Light of Kuiper Belt Fluxes: Evidence for Small Kuiper Belt Objects and Recent Geological Activity, *Astron. J.* 119: 945–952.

Thomas, P. C. (2000). NOTE: The Shape of Triton from Limb Profiles, *Icarus* 148: 587–588.

Tremaine, S., Touma, J. & Namouni, F. (2009). Satellite Dynamics on the Laplace Surface, *Astron. J.* 137: 3706–3717.

Tsurutani, B. T., Miner, E. D. & Collins, S. A. (1990). A close-up view of Triton, *Earth and Space* 3: 10–14.

Vokrouhlický, D., Nesvorný, D. & Levison, H. F. (2008). Irregular Satellite Capture by Exchange Reactions, *Astron. J.* 136: 1463–1476.

Ward, W. R. & Hamilton, D. P. (2004). Tilting Saturn. I. Analytic Model, *Astron. J.* 128: 2501–2509.

Yoder, C. F. (1979). How tidal heating in Io drives the Galilean orbital resonance locks, *Nature* 279: 767–770.

Zhang, K. & Hamilton, D. P. (2008). Orbital resonances in the inner neptunian system. II. Resonant history of Proteus, Larissa, Galatea, and Despina, *Icarus* 193: 267–282.

6

Enigma of the Birth and Evolution of Solar Systems May Be Solved by Invoking Planetary-Satellite Dynamics

Bijay Sharma

Electronics and Communication Department, National Institute of Technology, Patna
India

1. Introduction

From ancient times there has been a quest to understand the position of human kind in the cosmic order and to develop predictive system which could warn us of the impending natural calamity. In a continuing quest for an accurate predictive system, in Greek times Ptolemy kept our Planet at the center of the Universe and propagated the Geo-centric World View [Gale (2005-2006), Lawson (2004)]. In 16th century at the height of Renaissance, in a paradigm shift work but which was very much in keeping with common-sense , Nicolaus Copernicus, mathematician, astronomer and catholic monk, presented his book *"De revolutionibus orbium coelestium* (on the Revolution of the Heavenly Spheres)" first printed in 1542 in Nuremberg, Holy Roman Empire of the German Nation[Hawking (2005), Kuhn (1957), Windleband (1958), Crowe (1990)]. It offered a new framework for calculating the positions of the planets and this computational framework was tied to a Helio-centric World View [Hawking (2005)].

This Helio-centric Model was a natural consequence of common sense logic because the Sun was the heaviest object. The mass of Sun had been established during the renaissance by Sir Issac Newton [Hawking (2005)].This simple model at one stroke removed all the anomalies observed in the motion of the planets till then. But still it stood against a wall. The concept of helio-centrism was very much there in Greek Times [Gomez 2011] but the religious dogma and over-possessiveness of the idea of superiority of human-kind over all living kinds compelled geo-centric world view as the correct and the official tenant of the Greek times.

This dogma persisted. Such were the dogmatism of the Dark Mediaeval Period that in 1553 Michael Servetus [Goldstone & Goldstone(2002), Janz (1953)] was burnt at stake for advancing new ideas contrary to those of the Church. New ideas were considered heretical ideas.

In 1584 a young theologioan and naturalist by the name of Giordano Bruno [Singer (1950), Yates (1964), Brix (1998)] came on the European Scene. He boldly proclaimed the correctness of Helio-contric Model and he went a step forward saying that all stars were like our Sun, that there may be many more *extra terrestrial solar systems , many more exo-planets and many more extra terrestrial intelligence*. There was nothing sacrosanct about Man and his Earth just as there is nothing special about Chinese Civilization and their Middle Kingdom. This was the final nail in his coffin.

In 1592 Bruno was arrested by the Inquisition, a Church Court. His philosophical and political views were censored and he himself was burnt at stake in 1600. He was the martyr of "Free Thought and Modern Scientific Ideas". He was the bold harbinger of a New Cosmology during the Italian Renaissance.

De Revolutionibus was banned "until corrected". In 1620 nine sentences were deleted and then it was brought into circulation.

The debate about **extraterrestrial intelligence** continued and it was argued that if indeed there is **extraterrestrial intelligence** elsewhere there must be Earth-like planets in our Milky Way Galaxy. It was also argued that SETI must concentrate in those regions of our Galaxy where Earth-like planets are most likely to be found by anthromorphic principles. By anthromorphic principles the best places to find life in our galaxy could be on planets that orbit the Red Dwarf Star. Gliese 876 falls in this category. It is one-third the mass of our Sun and only 15 light years distant from us. It is three planet system. The planets falling in "Goldilocks Zone" around these Red Dwarfs will have maximum probability of **extraterrestrial intelligence**. These zones are the area around the star which is neither hot nor cold for liquid water to stay. The full lifecycle of a star is dependent on its mass. The lifecycle is inversely proportional to the mass. The massive stars are short lived, their life being of million years. The light stars like Red Dwarf star are very long lived, their life cycle extend up to 100 billion years. Therefore Red Dwarf planetary system has a greatest chance of harboring an evolved form of life. Thus the idea of Extra-Solar Systems and Exo-Planets were born. Extra-Solar Systems are the Solar –Systems around other main-sequence stars and members of the extra solar –systems are exo-planets.

M Dwarf or Red dwarf stars are most abundant outnumbering sun-like G Type stars by 10 to 1. Since these stars are likely to have earth like planets falling in Goldilocks Zone hence they are the primary target for SETI missions.

The following table gives the types of Stars and the likelihood of finding extra-solar systems:

Types	Mass	Likelihood
F- Type	1.3 to 1.5 M_Θ	10%
G- Type (sun like)	1 M_Θ	7%
K-Type	0.3 to 0.7 M_Θ	3 to 4%
M-Type	0.1 to 0.3 M_Θ	Unlikely.

Table 1. The types of stars and the likelihood of extra-solar systems with different types. [Zimmerman 2004]

2. The discovery of first extra-solar system[1]

In 1986, two proposals came from the University of Arizona and the University of Perkin-Elmer for space based direct imaging of Extra-Solar Systems using 16m- infrared telescope and optical telescope respectively.[Shiga 2004, Zimmerman 2004]. Atmospheric turbulence smears the star's light into an arcsecond blob and reduces the resolution therefore ground based imaging of exo-planets was impossible.

---

[1] [Lissauer 2002]

Adaptive Optics overcomes the atmospheric turbulence. Adaptive optics measures the scrambling due to air turbulence with a special sensor, then sends the information to a flexible mirror that deforms and undulates many times a second to tidy up the image. The rapid changes in the shape of the mirror exactly compensates the distorting effect of the churning atmosphere.

Recently extreme adaptive optics has been developed. It replaces hundreds of tiny pistons that reshape current flexible mirrors with thousands of smaller ones, and correct the incoming light not hundreds but thousands of times a second. This would spot a young glowing Jupiter in a much wider orbits. The road to another earth lies through another Jupiter, hence the presence of wide orbit Jupiter will mark the stars which should be closely examined first for earth like planets and then for life and intelligence.

In 1991 the first extra-solar system around a Pulsar was discovered by Alexander Wolszczan and Dale Frail. This pulsar is PSR1257+12, a rapidly rotating neutron star about $1.4M_\odot$ and at a distance of 2000 to 3000 light years of our Earth. In this solar-system three planets were observed. The two planets have orbital period of a few months, small eccentricities and masses a few times as large as the mass of Earth. Third planet, innermost planet, has a period of one month and the mass is that of our Moon.

Name	Jupiter	Gliese 229B	Teide1	Gliese229A	SUN
Type of object	Planet Gas Giant	Failed star Brown Dwarf	Failed star Brown Dwarf	M type Main Sequence Star Red Dwarf	G type Main Sequence Star, Yellow Dwarf
Mass($\times M_J$)	1	30	55	300	1,000
Radius(km)	71,500	65,000	150,000	250,000	696,000
Temperature(k)	100	1,000	2,600	3,400	5,800
Age(years)	4.5Gy	2-4Gy	120My	2-4Gy	4.5Gy
Hydrogen fusion	No	No	No	Yes	Yes
Deuterium fusion	No	Yes	Yes	Yes	Yes
Distinguishing feature of star.	No fusion whatsoever	Not hot enough for Hydrogen Fusion but deuterium fusion starts and after that the fusion fizzles out. Hence we say it is a failed star.	Not hot enough for Hydrogen Fusion but deuterium fusion starts and after that the fusion fizzles out. Hence we say it is a failed star.	Full scale fusion takes place from Hydrogen onward till Iron is nucleosynthesized. It can't go beyond Iron since Iron has the maximum binding energy.	Full scale fusion takes place from Hydrogen onward till Iron is nucleosynthesized. It can't go beyond Iron since Iron has the maximum binding energy.

Table 2. Distinction among Planets, Brown Dwarfs and Main Sequence Stars.

In 1994, 60-inch telescope on Palomar Mountain, coupled with primitive adaptive-optics system, imaged a brown dwarf orbiting the star Gliese 229. The brown dwarf was orbiting the host star at a semi-major axis of 40AU(Astronomical Unit) where 1AU is 1.5×10^8 km. The same system was photographed by Hubble Space Telescope. The ground-based imaging of this binary-star was confirmed by space image. This established the technical feasibility of taking ground-based images of sub-stellar objects using telescopes fitted with adaptive-optics.

In 1995 Mayor and Quiloz discovered the first exo-planet orbiting the star 51Pegasi. They used ELODIE spectrograph. In this the wobbling motion of the host star is used to detect the companion object. The wobbling motion of the host star gives rise to an effective radial velocity along the line-of-sight. Hence light coming from the host star experiences Doppler Effect. When the host star is approaching us , we record a blue shifted light and when host star is receding we record a blue shifted light. The recording of the alternate blue and red shift along the time axis gives the orbital period of the exo-planet and the magnitude of the shift gives us the mass of the host star. Since we may not be having an edge-on view of the orbital plane and the orientation radius vector of the orbital plane may be at an angle i, the angle of inclination of the orientation vector with respect to the line-of-sight, therefore the mass observed is MSini. We do not get the true mass of the exo-planet unless we have an edge-on view.

In 51Pegasi extra-solar system, we have the exo-planet orbiting the host star at a semi-major axis of 4.8 million miles. The orbital period is 4.2 days. This exo-planet is named 51 Pegasi.b. The mass observed, i.e. MSini , was more massive than that of Saturn.

One of the biggest drawback of Doppler Method of detection is that only Gas Giants of the size of Jupiter and Saturn can be detected.

ELODIE spectrograph has been further improved into CORALIE echlie spectrograph mounted on the 1.2m-Euler Swiss telescope at La Silla Observatory, ESO, Chile. This has been refined and exo-planets of Uranus mass have also been detected.

In 1999, a planet around HD209458 was detected by transit method. The actual mass and the size of the planet orbiting HD 209458 has been determined by combining the transit method and Doppler shift method. The density has been inferred and it is established that HD 209458b is a gas giant primarily constituted of Hydrogen just as Jupiter and Saturn are.

In 2001 the exoplanet OGLE-TR-56b detected by transit method. A polish team using 1.3m Warsaw Telescope at the Las Campanas Observatory in Chile made this discovery. In the transit method a dip in star light is caused while the exo-planet is transiting across the host star just as we record a solar eclipse when Moon is transiting across the face of Sun on NO MOON day. In the case of OGLE-TR-56b the dip occurred for 108 minutes and repeated every 1.2 days. Using 10m Keck I telescope on Mount Kea, Hawaii, the finding was confirmed by Doppler Method in January, 2003.

Both these discoveries were too close to the host star for comfort. In the classical model there was no place for gas giants to be orbiting closer than 1 to 2 AU. These exo-planets were called hot-jupiters and they defied the conventional wisdom.

3. The menagrie of exo-planets discovered till date[2]

708 exo-planets have been discovered till 17th December, 2011. 81 multiple exo-planet systems have been discovered till now. 10 earth and super-earths discovered. 2 of these are in Goldilock zone.["Coming Soon, Earth's Twin." The Economic Times on Sunday. December 11-17, 2011 Pg.15.] Generally the exo-planets have eccentricities equal to zero. That is they are orbiting in perfect circular orbits like our nine planets. But there are other exo-planets which are in highly elliptical orbits like comets. Planets have been found orbiting binary stars, in circum-binary configuration, as well as in three star-systems. [Doyle et. al. (2011), Welsh et.al(2012)]. Planets have been found orbiting pulsars.

The only exoplanet with an orbital period larger than that of Jupiter is the one orbiting 55Cancri. Its MSini = $4M_J$ and its orbital period is 14 years.

Planet as massive as $14M_E$ have been discovered around Mu Arae [Appenzellar 2004].Orbital period is 9.5 days. Hence it is very close to the parent star.

1992 Arecibo Radio Telescope	*Scientists announce the discovery of planets around a pulsar – a spinning neutron star. They are unlike any known planets and almost certainly hostile to life but are the first exo-planets to be found.*
1995 Haute –Provence Observatory	*Astronomers discover a planet around a sunlike star, 51 Pegasi, by tracking stellar motions. This is the Doppler Shift method. The same technique has revealed more than 130 planets.*
1999 STARE Project.	*For the first time the shadow of a Jupiter-size planet is detected as the planet passes across the face of the star , HD 209458. This is the transit method.*
2001 Hubble Space Telescope.	*By observing light from HD 209458 as its planet passes, astronomers see hints of a planetary atmosphere containing sodium.*
2003 Keck Interferometer	*The interferometer combines light from two existing Keck telescopes, eliminating atmospheric " noise" with adaptive optics. It will search for debris disk around stars, which could signal planet formation, and look directly for giant planets.*
2006 Large Binocular Telescope	*Its twin mirrors will search for debris disk and for newly formed Jupiter-size planets.*
2007 Kepler Mission.	*This space-based telescope is surveying more than 100,000 stars for dimming that hints at the presence of Earth-size planets.*
2009 Space Interferometry Mission (SIM)	*SIM will combine light from multiple telescopes to map stars and seek planets almost as small as Earth.*
2014-2020 Terrestrial Planet Finder (TPF)	*A two part space mission, TPF will detect from Earth-size planets and search for signs of habitability.*
2025? Life Finder	*The space- based Life Finder will search newfound Earths for signs of biological activity.*

Table 3. Chronological Order of the milestones achieved in exo-planetary studies.[Appenzellar 2004]

---

[2] [Shiga 2004, Zimmerman 2004]

Planets have been orbiting very close to their parent star so much so that they are slowly evaporating due to the heat and solar wind from their parent star. These are the hot Jupiters referred to above. As mentioned these defy the common wisdom of planet formation. By the year 2000, dozen exoplanets discovered and majority of them were hot Jupiters.

The catalog of exoplanets is growing and hot-jupiters seem to be an exception. The average planet size is falling and orbital distance is growing. That is exo-planets are being discovered farther and farther away from their parent star.

Table 3. gives a chronological order of the milestones achieved in exo-planetary studies

4. Conditions conducive to exo-planet growth

In general it is found that single star system favour planet growth. Heavier stars favor giant planet growth while lighter stars favour terrestrial planet growth [Thommes et.al.(2008)]

The extra- solar systems have a much larger probability in younger and more metal-rich regions of the spiral galaxies. The parent stars of exo-planets have higher metallicity [Santos 2005]. They have a higher abundance of elements heavier than hydrogen and helium.

The time factor is also very important. There is a very narrow time slot of few million years after the birth of the solar nebula in which the planets can be formed. The building blocks of planets are dust and gas. The dust particles of the accretion disc are continuously spiraling into the parent star by Poynting-Robertson drag and gas-dust smaller than 0.1 micron are being pushed out by solar radiation insolation by the process known as photo-evaporation [Ardila 2004].

In our Solar System there exists dusty debris disk in the asteroid belt. This causes the zodiacal light hence it is called zodiacal belt of dusty debris. This extends from 3AU to 10AU. There also exists Kuiper Belt of dusty debris from 30AU to 100AU. Similar dusty debris disk surround the stars with planetary system. These have been imaged by IRAS(infra red astronomical satellites) in 1983.It carried out complete survey of the sky in mid to far infra-red wavelength from 12 to 100 microns. The star itself is too hot, about 1000 Kelvin, to emit at far IR. But an accompanying debris disk will heat up and reradiate at far IR. This will give a bump in the stellar spectrum. The excess energy at infrared wavelength invariably indicate the presence of dusty debris disk. These debris disks are tenuous and faint but they have definite IR hazy glow. A gap in the debris disk is the signature of a protoplanet orbiting the parent star. The planet is in formative stage.

The dust in the debris disk either comes from the collisions of the initial leftover planetismals during planet formation or could be coming from collisions of comets and asteroids much after the formation has been completed. This debris disk generally range from 100AU to 1000AU and their composition is similar to that of our comets. The central part is a gap.

Ground based detectors cannot observe   IR because of the absorption effect of the atmosphere. Milllimetric radiations reach the surface of the Earth. Therefore Submillimeter Common-User Bolometer Arrays (SCUBA) are used on the ground observatory for detecting the mm radiation coming from the debris disk of the stars. A combination of IR and mm wavelengths observations made by Hubble Space Telescope, SCUBA and IR detectors from the ground observatories have established that a dozen stars possess the dusty debris disk including Beta Pictoris. These debris disks are the analogue of Kuiper belt debris and hence are cooler than expected.

The debris disk depend on the age. Young stars in formative stage possess a much larger and heavier dusty debris disk as compared to our Solar System which is 4.56Gy.In our Solar System much of the debris has been used up in planet formation and the residual has spiraled in due to Poynting-Robertson(PR) drag or photoevaporated. The dust presently seen in asteroid belt and Kuiper belt is the result of collision and evaporation of comets and asteroids. They are continuously being removed by PR drag and by photoevporation and they are also being replenished by collisions and evaporation. Hence the young stars have a much larger debris disk.

So far the stars with debris disk have not given the confirmation of the presence of planets and stars with extra- solar systems have not shown up any debris disk.

Name of the star	Age	Extent of the dusty debris disk	implications
HD 100546	<500My	? Revealed a gap at 10AU	A protoplanet might be orbiting the parent star.
Beta Pictoris	15My	1400AU edge-on disk debris disk detected at optical and near IR.	10,000 times as much dust as our solar system does. This means it has 100 times more planetismals as compared to our sun.
HD 141569	< 20My	Long spiral arms of dust. Debris disk detected at optical and near IR.	The companion stars could have created these features. It could be due to accompanying planets.
Fomalhaut		200AU in radius , edge on ring of dust is observed. Debris disk detected at thermal IR. A ring of warm materials detected very near the star.	Observed at 70 microns by SPITZER. The inner warm ring is akin to asteroid belt and its IR glow was observed at 24 micron.
Au Mic (M Type star)	15My	50 AU to 210AU	Excess of far-IR radiation points to the existence circumstellar dust grains;
HR4967A	< 20My	debris disk detected at optical and near IR.	
Vega		debris disk detected at thermal IR.	
ε Eridani		debris disk detected at thermal IR.	

Table 4. Stars with dusty debris disk and the implications.

The debris disks have definite large scale features such as rings, warps , blobs and, in one case, a large spiral. All the extra-debris disks so far detected are much more massive than our Asteroid belt debris and Kuiper belt debris.

Till date(1.01.2012) in last 16 months, since the Kepler Program was started, 2,326 planet candidates have been discovered out of which 31 have been have been confirmed. Kepler 22b is orbiting Sun-like star whereas Gliese-581d and HD 85512b are orbiting smaller and cooler stars but they are all in Goldilock zone.

The discovery of earth-like exo-planet would be the Holy Grail of astrobiology- a place where life started from scratch independently of life on Earth. The strategy is to first detect an earth-like exoplanet in the Goldilock zone of some star nearby say within 100 lightyears and then use terrestrial planet finder (TPF) to detect the biomarkers in the atmosphere of the given exoplanet.

5. Evolution of solar system building material

In NASA's DEEP IMPACT mission a 820 pound impactor collided with Comet Tempel 1. By the study of Comet material it was concluded that it was made of the pristine constituents of early solar system. This pristine material consisted of fragile organic material. This material includes polycyclic aromatic hydrocarbons(carbon based molecules found on charred barbeque grills and automobile exhaust on Earth).

On the other hand, the asteroids are the leftovers of planet formation and they therefore represent a more evolved form of matter. About 4000 Asteroids have been categorized. The Asteroid belt exists from 2.1AU to 3.3AU. Asteroids are coplanar with Ecliptic Plane. They move in the same direction as the Planets.

In terrestrial planets there is a metallic core and surrounding basaltic-granitic mantle.

But a Solar System which is in transition like HD113766 and which has a dusty disk has material in between the primitive kind contained in comets and more evolved kind found in asteroids.

Planet bearing Extra Solar Systems invariably have an environment rich in metal[Santos 2005]. The stars with twice the metallicity have 25% chance of harbouring a planet whereas stars with Sun's metallicity has only 5% chance.

There is a very narrow time slot of tens of millions of year in which Gas Giants birth and growth must take place. The dust part is continuously spiraling inward due to Poynting - Robertson photo assisted drag and gas-dust particle smaller than 0.1 micron are pushed out by the solar radiation insolation also known as photo evaporation.. Thus the gas-dust circumstellar disc is dissipated after tens of millions of years. If the opportunity is not seized for the birth and evolution of Gas Giants then no planetary formation would take place. The formation of Gas Giants is essential for Earth- like terrestrial planets.

6. The difficulties in discovering exo-planets

Doppler shift technique is the most convenient method of detecting Jupiter sized planets in tight orbits around their parent stars. The other methods are enumerated in Table 5.

Method	Description
Terrestrial direct imaging	Largest telescopes such as Keck, Gemini and Subaru are being used for direct imaging. Orbital architecture can be determined hence true mass is known. Easier to detect gas giants in wide orbits like ours. Young stars are ideal target as their companion planet would be glowing brightly in infra red wavelength because of the accretion generated heat.
Space direct imaging	William SPARKS(Space Telescope Science Institute) is using Hubble Space Telescope's Advanced Camera for Surveys for direct imaging. Orbital architecture can be determined hence true mass is known. Easier to detect gas giants in wide orbits like ours. Young stars are ideal target as their companion planet would be glowing brightly in infra red wavelength because of the accretion generated heat.
Radial Velocity technique or Radiovelocimetry or reflex motion of solar type stars	A color change in the star light betrays the wobble caused by the companion planet. When star is approaching, light experiences a blue shift and when star is receding, light experiences red shift. This is also known as Doppler Shift technique. There is uncertainty about the orbital angle of inclination hence real mass is indeterminent. Only the lower limit of the true mass is determined. Easier to detect gas giants in tight orbit.
Astrometric method	Recording the proper motion of the star on the celestial sphere i.e. the dome of the sky. Most sensitive for gas giants in nearby stars. Since 2-D picture is obtained therefore actual mass is determined. Wide orbit planets produce larger amplitude of the proper motion of stars hence easier to detect but wide orbit means longer orbital period hence a longer timeline of observations.
Transit photometry method	If the planet lies in the orbital plane of the star and we have an edge on view then the planet transit or Venus transit-like will cause a periodic square-well shaped dip in the star's brightness. It gives the estimate of planet size and the orbital period. The mass will have to be determined by astrometric or Doppler shift technique.
Gravitational microlensing	This method is used for detecting very faint stellar and sub-stellar bodies within our galaxy. A massive body intervening the space between the source and observer causes gravitational bending of light from the source leading to the brightening of the image of the source. If the intervening body is a star with a planet then the lining up of the source planet, intervening star and the observer will lead to considerable brightening up of the image of the source. As planet moves out of the line of sight, the brightening will diminish. The

Method	Description
	period of fluctuation in the image of the source is the orbital period and the amount of fluctuation gives the mass of the planet.
Lyot method	Suppresses 98.5% of the starlight by the use of a coronagraph and images the companion planet at near-IR wavelength or images the starlight reflected by the companion planet or by the circumstellar debris disk.
Nulling interferometry.	Large binocular telescope is used for canceling the starlight by nulling interference and image the exo-planet or the debris disk. Starlight are collected by two mirrors but with a path difference of half wavelength. This results into destructive interference along the central line of sight but it is constructive interference off the line of sight.
Radio emissions similar to those from Jupiter.	Radio emissions similar to those from Jupiter could reveal the presence of planets.

Table 5. Various methods of detecting exo-planets.[Shiga 2004],

Time is the greatest difficulties. The orbital periods of Jupiter and Saturn are 12 and 29.5 years. Hence one will have to wait for that long to measure its periodicity.

Second is the resolution of the Doppler Technique. With the present resolution we could keep looking for century and not detect a Saturn of that mass and of that semi-major axis.

The masses of Jupiter and Saturn are 318 and $95M_E$ and those of Neptune and Uranus are 17.2 and 14.6 M_E . The amplitude of Doppler oscillation is proportional to $(M\sin\alpha)/a^{1/2}$. Hence observational bias is towards heavier masses and shorter semi-major axis.

Table (6) gives the radial velocity which have been detected [Schwarzschild 2004].

Mass of the host star ($\times M_\odot$)	Mass of the planet ($\times M_E$)	Semi-major axis a (AU)	Amplitude of oscillation of Radial velocity of the host star (meter/second)
Red dwarf –Gliese 436 0.5	21	2.6 days 0.028AU	18
μ- Arae 1	14	9.5 days 0.084AU	4
ρ Cancri sun-like 1	18	2.8 days 0.04AU	6
Sun-like star 1	1	1AU	0.1
Sun	Jupiter	11.86 yrs 15AU	12.5
Sun	Saturn	30 yrs 20AU	2.7

Table 6. A comparative study of the radial velocity of the host star for different combinations of star-planets.

As seen from the Table(6), with decreasing mass of the planet and increasing mass of the host star, the amplitude of oscillation of the radial velocity decreases. As the amplitude of oscillation decreases it becomes increasingly difficult to decipher the periodic planetary signal in the presence of various noise sources that produce random fluctuations in star's apparent radial velocity.

Recently HARPS spectrometer has been developed by the Swiss team which discovered the first exoplanet Pegasi 51. This spectrometer has the required precision to decipher the tiny Doppler shift due to 0.1 m/s radial velocity of the host star harboring an earthlike planet. The instrument is kept in high vacuum and precisely controlled low temperature so that the sources of noise can be eliminated and optical stability may be imparted for obtaining the required precision. Through HARPS only Mu Arae's planet, 14 times M_E, was detected.

By Astrometric measurements , the inclination of the planetary orbit and orbit-globe parameters can be determined. Astrometry is the precise measurement of two-dimensional stellar positions on the celestial sphere. The astrometric studies complement the radial velocity method. Through this method the ellipse traced by the centroid of the star during one orbital period of the planet can be exactly determined. From this ellipse the angle of inclination α and other globe-orbit parameters can be determined.

Space Interferometry Mission scheduled for 2009 will give sufficient accuracy to astrometric method for discovering a new planet.

7. The classical model of the birth and evolution of a solar system

From the three new Neptune-like planets [Schwarzschild 2004] the scientists conclude the following:

i.   The shock waves of a Supernova explosion sets a giant cloud of gas and dust , passing nearby, into a spinning mode. The rapid spin cannot be accommodated by one hydrostatic star hence it results into the fragmentation of the cloud into binary or multiple star system. Even the new multiple system cannot accommodate the excess angular momentum and the individual clouds are flattened out as pancake shaped disc of accretions. The central part collapses into a proto-star surrounded by a thick disk of gas and dust. From these Keplerian debris disks the planets are born. The solar insolation is causing the photoevaporation of gas out of the system and the dust particles are spiraling inward due to Poynting-Richardson Drag and settling down in the midplane of the disc. Thus gas is blown out and the host star vicinity is filled with heavy suspension of dust particles larger than a micron size. These micron size dust randomly collide and stick together building up km-sized planetismals. But before the build up can take place the random collision may result in repeated breakups preventing the formation of planetismals. But if there is heavy dust suspension, with the gas blown out, runaway gravitational accretion takes place resulting into full scale terrestrial planets.

So there are two scenarios:

a.   The first scenario is the earliest stage of planet formation when the protostar is not experiencing full scale thermonuclear fusion . At that stage there is a very light density suspension of dust in a thick envelope of gas. The gravitation is too weak and gravitational accretion is prevented. But snowline criteria is not applicable as thermo-nuclear furnace is not switched on yet. Hence the dust is coated with ice which is amorphous and hence sticky (Ordinary ice is a open-pack hexagonal crystalline structure and is non-sticky whereas ice at -230°c is fluffy amorphous structure. If small ceramic ball is covered with fluffy, amorphous ice falling from a height of 12 cm it bounces to 1 cm whereas ball covered with crystalline ice bounces to 8 cm. The colder, more disordered ice absorbs more of the energy of the impact because the molecules rearrange themselves during the collision. Therefore the dust particles coated with amorphous ice will stick together rather than rebounce). Through collision and agglomeration (or sticking), km-sized planetismals are formed which are then set on the path of gravitational accretion. Once $10M_E$ cores have formed the gravitational field is strong enough to cause the wrapping of these icy-rocky cores with thick envelopes of gas resulting first into gas giants and subsequently into ice giants.

b.   The second scenario is when gas has been exhausted both by the process of gas giants and ice giants formation and also by photoevaporation. At this stage lack of gas assists runaway gravitational accretion of the thick dust suspension into terrestrial planets. Radioactive dating of the core by Hf-W has established that Earth and Mars were formed 29 million years and 13 million years respectively after the birth of the solar nebula [Cameron 2002, Yin et al 2002, Kleine et al 2002]. There was an extended core formation period. The interior of the planet is heated partly due to Helmholtz Contraction(or gravitational energy release) and partly due to radioactivity particularly that of[26]Al. Accumlative collision between small bodies produce the planet. When a small body collides into a large body the core of the small body gets embedded into the mantle of the large body. The heat of impact melts the interior and molten iron core of the smaller body percolates to the core of the larger body.

ii.   According to core-accretion theory or dust bunny theory, by agglomeration-accretion a rock or ice core is formed of mass 10 M_+ . Beyond that critical mass the core rapidly envelopes itself by gravitationally captured gas from the surrounding circumstellar disk. This process terminates with the formation of a gap in the circumstellar disc.

Douglas Lin(University of California, Santa Cruz) says " Many incipient gas giants won't make it to jovian mass before the disk dissipates after a few million years. So we can expect lots of failed Jupiters to show up as Neptune".

The farther the planet is the longer it takes to form. Infact it may be 100 billion years whereas the lifetime of the debris disk may be only several million years.

Computer models of Jonathen Lunine give the following picture:

•   In the inner part of the solar system debris disk is dense. In this dense part, the gas giants are formed in first million years through a chain of core formation and gas accretion;

•   In the next ten million years the leftover rock and dust accreted to form the moon – sized embryos. Dust clumps together into gravel, gravels to rock and rocks to hundred of planetary embryos moving in tidy, sedate circular orbits. The collisions stop.

•   Jupiter's influence that is gas giant's influence have two effects:

•   It churns an orderly set of embryos into an unruly, colliding swarm which through collision and accretion evolves into a set of terrestrial planets like our Earth and Mars in another 10 to 20 million years but these rocky planets are bone dry;

- Gas Giants in outer regions would cause icy embryos to veer inward and collide with newly evolved rocky planets. In the process water is transferred to the inner rocky planet;
- Gas Giants also act as bodyguards for these small watery worlds. There are large chunks of residual rock an ice which are on the loose and which would smash the inner rocky planets in next 100 million years. Gas Giants with its powerful gravitational fields took direct hit from these marauder chunks, some were flung out of the system and most others were herded into the asteroid belt;

iii. According to gas instability theory there is an abrupt formation of gas giants. The gravitational   instability in the circumstellar disc leads to gas-giant formation. There is no unfinished middleweights planets.

iv. In classical theory the explanation given for the infernally tight orbits of the hot jupiters is the following:

These must have formed much farther away beyond the snow-line which is about 1AU. Subsequently the tidal interaction with the protoplanetary disc caused the hot Jupiter to spiral in. This protoplanetary disc itself dissipates off due to Poynting-Robertson drag and due to photo- evaporation. So the inward migration must be fast before the dust-gas protoplanetary disc dissipates off [Schwarzschild 2004]. This is too contrived a situation. But in the new planetary-satellite model this naturally occurs.

8. The extra-solar planets which donot fit in any model

Lately many exoplanets have been discovered apart from hot-jupiters which donot fit any Model of planet birth and evolution and hence present a conundrum. Table (7) presents the list of the exoplanets and the reasons why they have become an enigma.

Name of the extra-solar system	Description	Reason for enigma
'Pegasi' exoplanets	Gas Giants in 100 times smaller orbit as compared to the orbit of Jupiter and Saturn	Gas Giants can form only beyond snowline which is at 1AU. Then how come hot-jupiters are in orbits of a less than 1 AU ? Gliese 436- a=0.028AU Mu Arae - a = 0.084AU Rho Cancri-a=0.04AU
HD 188753 (triple star system)	Hot Jupiter orbiting the primary star; Orbital period=3.35d; Orbital radius = 0.05AU; Mass= 1.14M_J ; Primary star mass=1.06$M_\odot$; Secondary system is a binary system of total mass=1.63$M_\odot$; Orbital radius of secondary with respect to the primary= 12.3AU; Orbital period of the primary and	A close and massive secondary will truncate the circumstllar disk around the primary to a radius of 1.3AU and the disk will be heated to temperatures which will prohibit the formation of a gas giant;*

Name of the extra-solar system	Description	Reason for enigma
	secondary around each other is 25.7y;	
HD 41004 (binary stars)	Stars are orbiting each other at distances of 20AU Primary containing the exoplanet is as massive as 3 times or more as compared to the secondary	
GI 186(binary stars)	Stars are orbiting each other at distances of 20AU; Secondary star is a white dwarf; Primary containing the exoplanet is as massive as 3 times or more as compared to the secondary	How did the planet of G186A survive the violent changing phases of the white dwarf, post main sequence evolution of star? A white dwarf is a spent out main sequence star which expands into Red Giant and then shrinks into a White Dwarf.
γ Cephei(binary stars)	Stars are orbiting each other at distances of 20AU & an orbital period of 56y; Primary containing the exoplanet is as massive as 3 times or more as compared to the secondary; Companion planet is $MSini=1.7M_J$ Orbital radius= 2.13AU; Orbital period=906d;	
19 binary or multiple star systems are inhabitated by a planet	Massive short period planets are found in multiple star system	Five short period planets in multiple star system cannot be explained in a classical fashion. [Eggenberger et al 2003]

* Initially it was thought that Giant planets must have formed in colder region far from their parent stars. Icy nuggets act as seeds that accumulate enough dust to build up to a critical mass where by runaway accretion it is enveloped by a large mass of gas giving birth to gas giants. These icy nuggets can form only beyond snowline[Sasselov & Lecar 2000]. But in HD 188753 this could not have happened. This is because the secondary system of star pair would truncate the disk to 1.3AU leaving nothing beyond in the colder region that could nucleate and grow into a giant planet.

"Giant planets in circumstellar disks can migrate inward from their initial formation positions. Radial migration is caused by inward torques between the planet and disk, by outward torque between the planet and the spinning star and by outward torques due to Roche lobe overflow and consequent mass loss from the planet." [Trilling, Benz et al 1998]. Through numerical solutions it has been shown that taking all the torques into consideration, Jupiter-mass planets can stably arrive and survive at small heliocentric distance just as hot-jupiters do in scorchingly tight orbits.

Table 7. The exoplanets which are conundrum.[Konacki 2005, Hatzes & Wuchterl 2005, Mugrauer 2005, Hatzes et al 2003]

9. Planetary satellite dynamics

On 21st July 1994, the Silver Jubilee Celebration Year of Man's landing on Moon, NASA gave a press release stating that Moon has receded by 1 meter in 25 years from 1969 to 1994. Using this piece of data, the first Author redid the analysis of Earth-Moon System [Sharma 1995]. In a subsequent paper the Authors [Sharma, B. K. and Ishwar, B " Basic Mechanics of Planet-Satellite Interaction with special reference to Earth-Moon System", 2004, http://arXiv.org/abs/0805.0100] found that Satellites-Planet Systems have a characteristic lom(length of month)/lod(length of day) equation:

$$LOM/LOD = E \times a^{3/2} - F \times a^2 \text{ [The proof is given in SOM\_Appendix A]}$$

Where l.o.m. = length of month (sidereal period of orbital rotation of the natural satellite around host planet which in case of our satellite Moon is 27.3 days);

l.o.d. = length of day (spin period of the host planet which in our case is 24 hours or 1 solar day);

a =   semi- major axis of the elliptical orbit of the satellite (for Moon it is 3,84,400 Km);

E = $J_T/(BC)$;

J_T =   total angular momentum of the Satellite- Planet System,

   = $(J_{spin})_{planet} + (J_{orbital})_{system} + (J_{spin})_{satellite}$;

B = $\sqrt{[G(M + m)]}$;

G =   Gravitational Constant = 6.67×10^{-11} N-m^2/Kg^2 ;

M =   mass of the host planet;

m =   mass of the satellite;

C =   Principal Moment of Inertia around the spin axis of the Planet;

F = $m/[C(1+m/M)]$;

When lom/lod = 1 we have geosynchronous orbit.

$$E \times a^{3/2} - F \times a^2 = 1 \tag{1}$$

Equation (1) has two roots and hence planet -satellite systems have two geo-synchronous. Only at these two Geo-synchronous orbits the system is in equilibrium because the orbits are non-dissipative. Elsewhere the system is dissipative hence in non-equilibrum either spiraling out to the outer geo-synchronous orbit or spiraling inward to its certain doom. The inner Geo-synchronous orbit lies at energy maxima whereas the outer Geosynchronous orbit is at energy minima. Therefore the inner geo-orbit is an unstable equilibrium orbit and the outer geo-orbit is a stable equilibrium orbit.

When the natural satellite is at the inner geo-orbit it is easily perturbed by solar wind or cosmic particles or solar insolation. It tumbles out on an expanding outward spiral path or it falls short of the inner geo-orbit on inward collapsing spiral path. Inward collapsing spiral path is entirely a runaway path. The outward spiral path, because of energy conservation, is initially an impulsive gravitational runaway phase which quickly terminates because of tidal dissipation in the central host body due to tidal streching and squeezing . This runaway phase is the gravitational sling shot phase. After the gravitational sling shot phase, the natural satellite coasts on its own towards the outer geo orbit. Our Moon is on a midway course in its journey towards the outer geo-orbit. Charon, a satellite of Pluto, has already arrived at the outer geo-orbit. The satellite may remain stay put in the outer geo-orbit as Charon is doing or it may be deflected as our Moon will be.

10. The new hypothesis- gravitational sling shot model of planet-satellite system

The Authors did the Keplerian-approximated analysis of Earth-Moon, Mars-Deimos-Phobos and Pluto-Charon [Sharma & Ishwar 2004A, Sharma, Rangesh & Ishwar 2009]. The Authors were able to generate the outward expanding spiral path of Moon as shown in Figure 1.

In a sequel paper on the New Perspective of the Solar System[Sharma & Ishwar 2004B, Sharma 2011], it was established that Planets experience a similar kind of impulsive sling-shot phase due to Sun as our Moon does due to Earth. This leads to new paradigm on the birth and evolution of our as well as extra Solar Systems.

10.1 THe phenomena of gravitational slingshot

Planet fly-by, gravity assist is routinely used to boost the mission spacecrafts to explore the far reaches of our solar system[Dukla, Cacioppo & Gangopadhyaya 2004, Jones 2005, Epstein 2005, Cook 2005]. Voyager I and II used the boost provided by Jupiter to reach Uranus and Neptune. Cassini has utilized 4 such assists to reach Saturn.

A space-craft which passes " behind" the moon gets an increase in its velocity(and orbital energy) relative to the primary body. In effect the primary body launches the space craft on an outward spiral path. If the spacecraft flies "infront" of a moon, the speed and the orbital

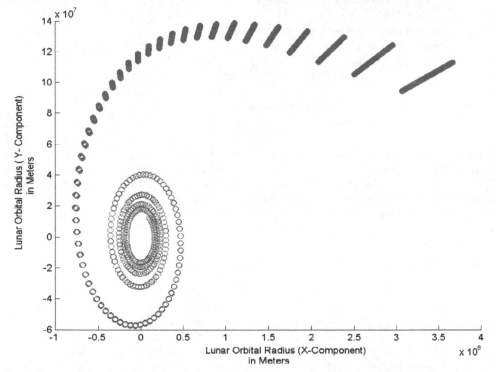

Fig. 1. Lunar Orbital Radius expanding spiral trajectory obtained from the simulation for the age of Moon (i.e. from the time of Giant Impact to the present times covering a time span of 4.5Gyrs).

energy decreases. Traveling "above" and "below" a moon alters the direction modifying only the orientation (and angular momentum magnitude). Intermediate flyby orientation change both energy and angular momentum. Accompanying these actions there are reciprocal reactions in the corresponding moon.

The above slingshot effect is in a three body problem. In a three body problem , the heaviest body is the primary body. With respect to the primary body the secondary system of two bodies are analyzed.

In case of planet flyby, planet is the primary body and the moon- spacecraft constitute the secondary system.

While analyzing the planetary satellites, Sun is the primary body and planet-satellite is the secondary system. But in our Keplerian approximate analysis, Sun has been neglected without any loss of generality and without any loss of accuracy. In fact the general trend of evolution of our Moon has been correctly analyzed [Sharma, B. K. and Ishwar, B " Basic Mechanics of Planet-Satellite Interaction with special reference to Earth-Moon System", 2004, http://arXiv.org/abs/0805.0100].

While analyzing the Sun-planet system, galactic center is the primary body and Sun-planet is the secondary system. But in our analysis the galactic center has been neglected and we have essentially analyzed Sun-planet as a two body problem.

In a similar fashion in the analysis of Planet Flyby-Gravity Assist Maneuvers, Planet is the primary body. The planet can be neglected and moon-spacecraft can be treated as a two body problem and the same results can be obtained without any loss of accuracy or generality. This will be done in a future paper.

The gravitational sling shot becomes clearer if we look at the radial acceleration and radial velocity profile.

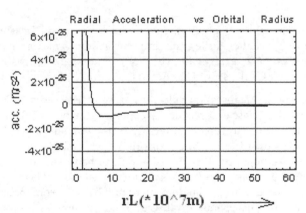

Fig. 2. Radial Acceleration Profile of Moon (Within a_{G1} the Moon is accelerated inward. Beyond a_{G1} the Moon is rapidly accelerated outward under the influence of an impulsive gravitational torque due to rapid transfer of spin rotational energy. The maxima of the outward radial acceleration occurs at a_1. (This is the peak of the impulsive sling shot torque.)

10.2 Setting up of the time integral equation.

In setting up the time integral equation the first step is to set up the radial velocity expression which has been derived in SOM_Appendix A.

The radial velocity expression is as follows:

$$\frac{da_{Iap}}{dt} = \frac{K}{a_{Iap}^M}\left(\frac{\omega}{\Omega} - 1\right) \cdot \frac{2a^{1/2}}{m \cdot B} = \frac{K}{a_{Iap}^M}\left(E \cdot a_{Iap}^{3/2} - F \cdot a_{Iap}^2 - 1\right) \cdot \frac{2a^{1/2}}{m \cdot B}$$

Or

$$v\left(a_{Iap}\right) = \frac{da_{Iap}}{dt} = \frac{2K}{a_{Iap}^M} \cdot \frac{1}{m \cdot B} \cdot \left(E \cdot a_{Iap}^2 - F \cdot a_{Iap}^{2.5} - \sqrt{a_{iap}}\right) \tag{2}$$

Where K is the structure constant and M is the structure exponent. All the other symbols are defined as before. Equation 2 gives the radial velocity of natural Satellite Iapetus with respect to Saturn.

Between a_{G1} and a_{G2}, ω/Ω is greater than Unity hence radial velocity is positive and recessive.

At less than a_{G1}, ω/Ω is less than Unity hence radial velocity is negative and secondary approaches primary.

At greater than a_{G2}, ω/Ω is negative which is physically not possible in a prograde system hence system is untenable and it is a forbidden state.

Spin to Orbital velocity equation yields a root when it is in second mean motion resonance (MMR) position. That is:

$$\frac{\omega}{\Omega} = E \cdot a_{Iap}^{3/2} - F \cdot a_{Iap}^2 = 2 \tag{3}$$

This gives a root at a_2 which is gravitation resonance point and I assume that after the secondary undergoes gravitational sling shot impetus, it attains maximum recession velocity at this point. After this maxima, recession velocity continuously decreases until it reaches zero magnitude at outer Clarke's Orbit as shown in Figure 3.

Thus as is evident from Eq.2, recession velocity is zero at a_{G1} and a_{G2}. From a_{G1} to a_2, the system is in conservative phase and secondary experiences a powerful sling-shot impulsive torque which imparts sufficient rotational energy to the secondary by virtue of which the secondary coasts on its own from a_2 to a_{G2} during which time the system is in dissipative phase, Secondary is exerting a tidal drag on the central body and all the rotational energy released by the central body as a result of *de-spinning* is lost as tidal heat, but not completely. This tidal heat is produced during tidal deformation of both the components of the binary if the secondary is not in synchronous orbit. Our Moon is presently in synchronous orbit hence it is not undergoing tidal heating but Earth is undergoing tidal heating.

When the secondary tumbles into sub-synchronous orbit it experiences a negative radial velocity which launches it on a collapsing spiral and the system is *spun-up* . In this collapsing phase, secondary exerts an accelerating tidal torque on the central body and

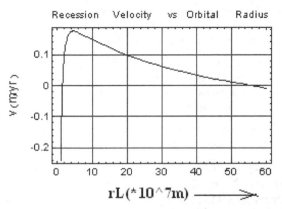

$$rL(* 10^{\wedge}7m) \longrightarrow$$

Fig. 3. Radial Velocity Profile of Moon. (Beyond a_{G1}, Moon is rapidly accelerated to a maximum radial velocity, V_{max}, at a_2 where Sling-Shot Effect terminates and radial acceleration is zero. Then onward Moon coasts on it own towards the outer Geo-Synchronous Orbit a_{G2})

rotational energy is transferred to the primary. This rotational energy causes spin-up of the central body as well as it tidally heats up the central body by tidal deformations.

Since Eq.2 has a maxima at a_2 therefore the first derivative of Eq. 2 has a zero at a_2. Equating the first derivative of Eq.2 to zero we get:

$$E(2-M)a_{lap}^{1.5} - F(2.5-M)a_{lap}^2 - (0.5-M) = 0 \ at \ a_2 \qquad (4)$$

From Eq.4, structure exponent 'M' is calculated.

We donot yet know the structure constant K. We make an intelligent guess of V_{max} and calculate the value of 'K' from Eq. 2 equated to V_{max} at semi-major axis 'a_2'.

Using these values of 'K' and 'M' the time integral equation is set up and tested for the age of the system.

$$\int \left[\frac{1}{v(a_{lap})} da, a_{G1}, a_{lapresent} \right] = transit \ time \ from \ a_{G1} \ to \ the \ present \ value \ of \ a_{lap} \qquad (5)$$

This transit time should be of the order of 4.5Gy in the case of Iapetus because that is the age of Iapetus.[Castillo- Rogez et al (2007)]. Through several iterations we arrive at the correct value of K.

10.3 Theoretical verification of the experimentally observed 'lengthening of day' curve of our planet Earth by primary-centric analysis[3]

Since the birth of Earth-Moon System, Earth's spin has been slowing down and Moon has been receding. Earth's spin has slowed down from 5 hours to 24 hours today and Moon has receded from 15,000Km to the present Lunar Orbit of 384,400Km.

---

[3] [Sharma, B. K. and Ishwar, B " Basic Mechanics of Planet-Satellite Interaction with special reference to Earth-Moon System", 2008, http://arXiv.org/abs/0805.0100].

John West Wells through the study of daily and annual bands of Coral fossils and other marine creaturs in bygone era has obtained ten length of day of bygone eras [Wells 1963, Wells 1966]. These benchmarks are tabulated in Table (8).

Leschiuta & Tavella [Leschitua & Tavella 2001] have given the estimate of the synodic month. From the synodic month we can estimate the length of the Solar Day as given in SOM_Appendix [C]. The results are tabulated in Table (9). [Leschitua & Tavella 2001 based on the study of marine creature fossils]

Kaula & Harris [1975] have determined the synodic month through the studies of marine creatures. The results are tabulated in Table (10).

One benchmark has been provided by Charles P. Sonnett et al through the study of tidalies in ancient canals and estuaries [Sonett & Chan 1998]. He gives an estimate of $T_{E4} = 18.9$ hours mean solar day length at about 900 million years B.P. in Proterozoic Eon, pre-Cambrian Age.

T (yrs B.P.)	T* (yrs after the Giant Impact)	Length of obs. Solar Day T_E^* (hrs)
65 Ma	4.46456G	23.627
135 Ma	4.39456G	23.25
180 Ma	4.34956G	23.0074
230 Ma	4.29956G	22.7684
280 Ma	4.24956G	22.4765
345 Ma	4.18456G	22.136
380 Ma	4.14956G	21.9
405 Ma	4.12456G	21.8
500 Ma	4.02956G	21.27
600 Ma	3.92956 G	20.674

Table 8. The Observed lod based on the study of Coral Fossils.

T (yrs. B.P.)	T* (yrs. After the Giant Impact)	Observed Synodic Month (modern days)	Estimated Solar Day (hrs).
900 Ma (Proterozoic)	3.62956G	25.0	19.2
600Ma (Proterozoic)	3.92956G	26.2	20.7
300Ma (Carboniferous)	4.22956G	28.7	22.3
0 (Neozoic)	4.52956G	29.5	24

Table 9. Observed Synodic Month

T (yrs. B.P.)	T* (yrs. After the Giant Impact)	Observed Synodic Month (modern days)	Estimated Solar Day (hrs).
45 Ma	4.48456G	29.1	23.566
2.8 Ga	1.72956G	17	13.67 (with modern C) 16.86 (with C = 9.99* 10^{37} $kg - m^2$)

Table 10. Observed Synodic Month (Kaula & Harris 1975) based on the studies of Marine creatures.

10.4 Comparative study of lengthening of day curve of our Earth by theory and observation

As seen from the superposition of the two lengthening of day curves, there is remarkable match between Observation and Theory in the recent past after the Pre-Cambrian Explosion

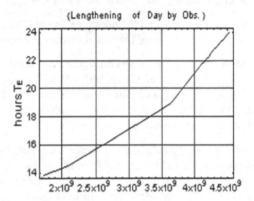

Fig. 4. Lengthening of Day Curve w.r.t. time by Observation

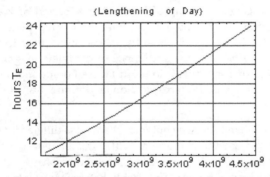

Fig. 5. Lengthening of day curve w.r.t. time by Theory assuming constant C.

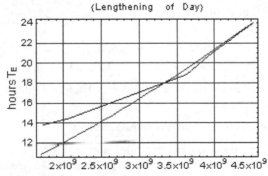

Fig. 6. Superposition of the two curves, one by observation and the other by calculation, with constant C.

of plant and animal life but in the remote past, particularly in early Archean Eon, Earth seems to be spinning much slower than predicted by theory. This implies that rotational inertia was much higher than what has been assumed in this analysis. In fact there are evidence to show that early Earth was much less stratified as compared to modern Earth. It was more like Venus [Allegre, Calnde 1994, Taylor, Rose & Mclennan 1996].

Through out the analysis C, the Principal Moment of Inertia, has been assumed to be constant whereas infact it was evolving since the Giant Impact [Runcorn 1966].

In the first phase of planet formation, Earth was an undifferentiated mass of gas, rocks and metals much like Venus. At the point of Giant Impact, the impactor caused a massive heating which led to melting and magmatic formation of total Earth. The heavier metals, Iron and Nickel, settled down to the metallic core and lighter rocky materials formed the mantle. The mantle consisted of Basalt and Sodium rich Granite.

Due to Giant Impact, Earth gained extra angular momentum. This led to a very short spin period of 5 hours. It has been calculated that the oblateness at the inception must have been 1% [SOM_Appendix D, Kamble 1966] whereas the modern oblateness is 0.3%. Taking these two factors into account C of Earth must have been much higher than the modern value of $8.02 \cdot 10^{37} \ kg - m^2$. In this paper the early C has been taken as $9.9 \cdot 10^{37} \ kg - m^2$.

After Achaean Eon the general cooling of Earth over a period of 2 billion years led to slower plate-tectonic movement. The 100 continental-oceanic plates coalesced into 12 plates initially and into 13 plates subsequently. The slower plate tectonic engine led to deep recycling of the continental crust and hence to complete magmatic distillation and differentiation of the internal structure into multi-layered onion like structure. Thus at the boundary of Archean Eon and Proterozoic Eon a definite transition occurred in the internal structure.

Before this boundary, the mantle and the outer crust was less differentiated. It was composed of a mixture of Basalt and Sodium-rich granite. After this boundary a slower plate-tectonic dynamo helped create the onion-like internal structure with sharply differentiated basaltic mantle and potassium-rich granitic crust. This highly heterogenous internal structure and less oblate geometry leads to the modern value of C equal to $8.02 \cdot 10^{37} \ kg - m^2$.

The form the evolving C is as follows:

$$f[(t-2E9)\_]:=If[(t-2E9)>0,1,0]$$
$$\{9.9E37-(9.9E37-8.02E37)\}\{1-Exp[-t/16E9]\}-f[(t-2E9)\_](1.4E37)\{1-Exp[-t/(0.5E9)]\}\} \tag{6}$$

Here f[(t-2E9)_] is defined as a step function which is 0 before 2 billion years and is Unity at 2 billion years and at greater times.

The profile of evolution of C with time is obtained in Fig. (7) :

As can be seen in Fig. (9), there is a much closer fit except for a large deviation at 2.5Gyrs after the Giant impact. This is due to step change in Moment of Inertia, C, at 2Gyrs after the Giant Impact. It would have been more realistic to assume a gradual change in C at the boundary of Archean and Proterozoic Eon. This correction will be made in a sequel paper.

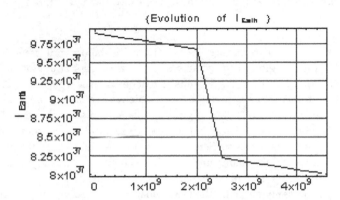

Fig. 7. The profile of the assumed evolving C.

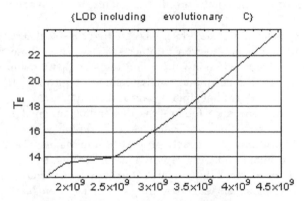

Fig. 8. Theoretical lengthening of day curve with evolving C.

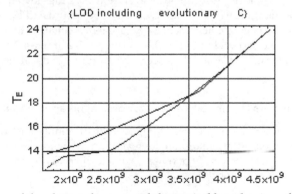

Fig. 9. Superposition of the observed curve and theoretical lengthening of day curve with evolving C.

10.5 A new perspective of birth and evolution of our solar system & extra solar systems

The new perspective holds that:

i. before the thermonuclear furnace turns on that is before full scale fusion reaction begins, in the inner region of the solar system by a process of agglomeration-accretion the icy-rocky core is formed. As soon as it reaches a critical mass of $10M_E$, it rapidly wraps itself with Hydrogen and Helium gas which is available in abundance in the gas-dust debris . As it grows to $300M_E$, a gaping hole is formed in the disk. This paucity of gas terminates the runaway gas accretion. As we see, the necessity of a snow line does not arise as the inner region is sufficiently cold (100 kelvin) to keep the dust coated with amorphous ice which eliminates impact rebounce and permits agglomeration to take place unhindered to km size planetismal.

ii. by the above process sequentially the four jovian planets are born i.e. one after another. As the first Gas Giant is formed, because of initial slingshot effect, caused by our Sun, Jupiter spirals out and makes space for the formation of the next gas giant namely Saturn. As Saturn spirals out, the Ice Giants namely Neptune and Uranus are formed.

iii. Just as Jupiter spirals out to wide orbits, it is equally probable that the gas giant may be perturbed within the inner geo-orbit in other solar systems. Those tumbling short of inner geo-orbit get launched on inward collapsing spiral path doomed to their certain distruction. They become hot jupiters in scorchingly tight orbit. In course of planet discovery, many examples of hot jupiters have turned up.

iv. the planet formation sequence follows the descending order of mass. The heaviest (i.e.Jupiter) being born the first and the lightest (i.e. Neptune) the last;

v. the time factor of evolution is inversely proportional to the mass i.e. the massive giants evolve out of their initial orbit very rapidly whereas the lightest one remaining almost stay put. This implies that Jupiter spirals out of the maternity ward very rapidly whereas the terrestrial planets remain orbiting where they were born;

vi. in the first phase, Gas Giants and Ice Giants are formed when there is abundance of gas and dust. In course of birth and evolution of these massive planets the disk is dissipated of gas partly due to the accretion by the jovian cores and partly due to photoevaporation. The remnant disk is largely populated by planetismals. In the second phase the rocky planetismals gravitationally collapse to form the terrestrial planets in a sequence according to the descending order of mass. Earth was formed the first and Mercury the last. Pluto is a recently captured body. It has not been formed insitu.

11. Observational proofs in support of gravitational sling shot model

In recent days four observations strongly suggest that in remote past Jupiter and the gas giants may have experienced gravitational sling shot and they may have been launched on an outward spiral path just the way Moon has been launched or for that matter all planetary natural satellites have been launched.

a. 700 Hilda asteroids in elliptical orbit [Franklin et al 2004].The asteroid belt is populated with hundred thousands of rocky remnants leftover from planet formation. These are called asteroids and they lie between Mars and Jupiter orbit between a radii of 3AU to 10AU. Most of the asteroids are in near circular orbits. There are 700 odd asteroids known as Hilda which are in highly elliptical orbit and these eccentricities could have been imparted only by a migrating Jupiter set on an expanding spiral path. The migrating Jupiter first ejected some proto-Hilda asteroids out of the system and next elongated the orbits of the residual asteroids. The migrating Jupiter could have also set the planetary embryos on unruly chaotic paths which led to frequent collisions and accretion resulting into terrestrial rocky planets.

b. Through computer simulation studies [Tsiganis, Gomes, Morbidelli & Lavison 2005] it has been shown that our planetary system, with initial quasi-circular, coplanar orbits, would have evolved to the current orbital configurations provided Jupiter and Saturn crossed the 1:2 mean motion resonance (MMR). When the ratio of the orbital periods of Jupiter and Saturn is 1:2 it is the strongest resonance point. At all integer ratios resonance is obtained but the maximum is obtained at 1:2. The resonance crossings excite the orbital eccentricities and mutual orbital inclinations to the present values. Jupiter ,Saturn and Uranus have the present eccentricities of 6%, 9% and 8% respectively. The present mutual inclination of the orbital planes of Saturn, Uranus and Neptune take the maximum values of approximately 2° with respect to that of Jupiter. The simulation was started with the initial positions of Jupiter and Saturn at 5.45AU and 8AU respectively. 1:2MMR crossing occurs at 8.65AU. The present orbital semi major axes of Jupiter, Saturn, Uranus and Neptune are 10AU, 15AU, 19.3AU and 30AU respectively. This simulation reproduces all aspects of the orbits of the giant planets: existence of natural satellites, distribution of Jupiter's Trojans and the presence of main belt asteroids.

c. The presence of Jupiter's Trojans can be explained only by 1:2MMR crossing by Jupiter and Saturn[Morbidelli, Levison, Tsiganis and Gomes 2005]. These are asteroids which are in he same orbit as that of Jupiter but they are leading or lagging by 60° in their co-orbital motion.

d. The petrology record on our Moon suggests that a cataclysmic spike in the cratering rate occurred approximately 700 million years after the planets formed[Gomes, Levison, Tsiganis and Morbidelli 2005]. With the present evidence we assume the birth of our Solar Nebula at 4.56Gya. The formation of Gas Giants and Ice Giants was completed in first 5 millon years and Earth was completed in first 30 million years. This puts the date of completion of Giant Planets at 4.555Gya and the date of completion of the Terrestrial Planets particularly Earth at 30 million years after the solar nebula was born that is at 4.53Gya. At 4.53Gya, the Giant Impact occurred and from the impact generated circumterrestrial debris, Moon was born beyond Roche's Limit at 16,000Km orbital radius. By gravitational sling shot effect it was launched on an outward spiral path. Presently Moon is at the semi-major axis of 3,84,400Km with a recession velocity of 3.7cm/year. Towards the end of planet formation phase, the residual debris of the solar nebula was being rapidly sucked in or swept out of the system. This resulted in heavy meteoritic bombardment of all the big sub-stellar objects including our Moon. Through Apollo Mission studies it has been determined that there is a sharp increase in the bombardment rate and hence in the cratering rate around the period of 4.5 to 3.855Gya. From this it is concluded that there was a cataclysmic Late Heavy Bombardment of all big sub-stellar bodies, including our Moon, at about 700 My after the completion of formation of Jupiter and Saturn.

As the planet formation was completed , the gaseous circumsolar nebula was dissipated by gravity accretion and finally by photoevaporation. According to Tsiganis et al [2005], Jupiter and Saturn were born at 5.45AU and 8AU respectively where the orbital period ratio that P_S/ P_J was less than 2. According to them the resulting interaction with massive disk of residual planetismals Jupiter and Saturn spiraled out on diverging path crossing 1:2MMR(P_S/ P_J = 2) point at 8.65AU and today the ratio is little less than 2.5. At the 1:2MMR crossing due to gravitational resonance their orbits became eccentric. This abrupt transition temporarily destabilized the giant planets, leading to a short phase of close encounters among Saturn, Uranus and Neptune. As a result of these encounters, and of the interactions of the ice giants with the disk, Uranus and Neptune reached their current heliocentric distances of 19.3AU and 30AU. And Jupiter and Saturn evolved to the current orbital eccentricities of 6% and 9%. The same planetary evolution can explain LHB provided Jupiter and Saturn crossed 1:2MMR 700My after their formation. That is LHB occurred at 3.855Gya.

12. Future direction of investigation

This new perspective on Solar System birth and evolution based on planetary satellite dynamics is called Primary-centric World View. This Primary-centric View has led to the fractal Architecture of the Universe [Sharma 2012A]. The Primary-centric View has been applied to Kepler-16b, Kepler 34b and Kepler 35b to explain its circum-binary architecture [Sharma 2012B]. The Primary-centric View has also been used to test the validity Iapetus's hypothetical sub-satellite [Sharma 2012C]. The Author has utilized the primary-centric view to work out the probable evolutionary history of PSR J1719 -1438 and its compacted companion at a distance of 4000ly[Sharma 2012D)]. Presently I am applying this World View to see if star binaries, pulsar binaries, pulsar-black hole, Galaxy of Stars, Clusters and Super-Clusters fall in this frame work or not[Sharma 2011]. A positive test for all these binaries and galaxy of stars will give us a new approach to the dynamics of the Universe.

13. References

Allegre, Calnde J. & Schneider, Stephen H. "The Evolution of Earth," *Scientific American*, October, 1994.

Appenzellar, Tim "Search for other Planets", *National Geographic*, December 2004.pp.75-93.

Ardila, David R. " Does our solar system represent the rule or the exception", *Scientific American*, April 2004, pp.36-41;

Basri, Gibor " The discovery of Brown Dwarfs", *Scientific American*, April 2000, pp 80-83;

Blue, Jennfer(9th November 2009) " Planets and Satellites Names and Discoverers", *USGS*,Retrieved 13th January 2010

Brix, H. James, *"Giordano Bruno"*, HARBINGER, Mobile, Alabama (1998)

Cameron, A. G. W. "Birth of a Solar System", *Nature*, Vol. 418, pp. 924-925, 29 August 2002.

Canup, R. N. & Esposito, "Origin of Moon in a giant impact near the end of the Earth's formation," *Nature*, 412, 16th August,2001.

Castillo- Rogez, T.C., Matson, D.L., Sotin, C., Johnson, T.V., Lunine, J.I. and Thomas, P.C. "Iapetus' geophysics: Rotation Rate, shape and equatorial ridge", *ICARUS*, 190, pp. 179-202, (2007);

Cook, C.L. "Comment on 'Gravitational Slingshot,'by Dukla, J.J., Cacioppo, R., & Gangopadhyaya, A. [American Journal of Physics, 72(5), pp 619-621,(2004)] *American Journal of Physics*, 73(4), pp 363, April, 2005.

Copernicus, Nicolus "De Revolutionibus Orbium Coelestium", 1933

Crowe, Michael J. *"Theories of the World from Antiquity to the Copernican Revolution"*, Dover Publications, Mineola, New York, (1990).

Doyle, Lawrance R., Carter, Joshua A., Fabrycky, Daniel C., et.al. "Kepler 16: A Transiting Circumbinary Planet", *Science*, 333, 1602-1606 (2011).

Dukla,J.J., Cacioppo, R., & Gangopadhyaya, A. " Gravitational slingshot", *American Journal of Physics*, 72(5), pp 619-621, May,2004.

Epstein, K.J. "Shortcut to the Slingshot Effect," *American Journal of Physics*, 73(4), pp 362, April, 2005.

Franklin, Fred A., Lewis, Nikole K., Soper, Paul R. and Holman, Mathew J. , " Hilda Asteroids as Possible Probes of Jovian Migration", *Astronomical Journal*, 128, 1391, 2004 September.

Galilei, Galileo "Siderus Nuncius", Thomam Baglionum, Venice (March 1610), pp.17-28

Gale, Thomson " Ptolemic Astronomy, Islamic Planetary Theory and Copernican's Debt to the Maragha School", *Science and its Times*, Thomson Corporation, (2005-2006).

Gomes, R., Levison, H.F., Tsiganis, K. and Morbidelli, A. " Origin of the cataclysmic Late Heavy Bombardment period of the terrestrial planets", *Nature*, Vol. 435, 26 May 2005, pp. 466-469;

Gomez, A. G., " Aristarchus of Samos, the Polymath", Journal of Scientific and Mathematical Research (Under Consideration for Publication) 2011.

Goldstone, Lawrence and Goldstone, Nancy "Out of the Flame: The Remarkable Story of the Fearless Scholar, a Fatal Heresy, and one of the Rarest Books", http://Amazon.com (2002)

Hawking, Stephen *"The Illustrated on the Shoulders of Giants"*, Running Press Publications, (2005)

Ida, S., Canup, R. M. & Stewart, G. R., "Lunar Accretion from an impact-generated disk," *Nature*, 389, 353-357. 25th Sept. 1997.

Janz, Daniel R.(eds) *"Verdict and Sentence for Michael Servetus"*, A Reformation Reader, 268-270, (1953).

Jones, J.B. "How does the slingshot effect work to change the orbit of spacecraft?" *Scientific American*, pp 116, November, 2005

Kaula, W. K. & Harris, A. "Dynamics of Lunar Origin and Orbital Evolution," *Review of Geophysics and Space Physics*, 13, 363, 1975.

Kerr, R. A. "The First Rocks Whisper of their Origins", *Science*, Vol. 298, pp. 350-351, 11 October 2002.

Kleine, T., Mezger, K., Palme, H., Scherer, E. & Munker, C. 'The W isotope evolution of the bulk silicate Earth: constraints on the timing and mechanisms of core formation and accretion,' *Earth Planet Science Letters*, 228, 109-123 (2004),

Kuhn,Thomas S. , " *The Copernican Revolution*", Harvard University Press, Cambridge, pp. 5 – 20, (1957).

Leschiutta, S. & Tavella P., "Reckoning Time, Longitude and The History of the Earth's Rotation, Using the Moon" *Earth, Moon and Planets*, 85-86 : 225-236, 2001.

Lissauer, Jack J. " Extrasolar planets", *Nature*, Vol. 419, 26 September 2002, pp. 355-358;

Maddox, J. "Future History of our Solar System", *Nature*, 372, pp.611, 15th Dec 1994.

Morbidelli, A., Levison, H.F., Tsiganis, K. and Gomes, R. " Chaotic capture of Jupiter's Trojan asteroids in early Solar System", *Nature*, Vol. 435, 26 May 2005, pp. 462-465;

Runcorn, S. K. "Change in the Moment of Inertia of the Earth as a result of a Growing Core", *Earth-Moon System* edited by Marsden & Cameron, Plenum Press, 1966., 82-92.

Santos, Numo C., Benz W. & Mayor M., "Extra-Solar Planets: Constraints for Planet Formation Model", *Science*, 310, 251-255, (2005),

Sachs, Abraham J. "Babylonian Observational Astronomy", *Philosophical Transactions of the Royal Society of London,*Vol.276(1257), 43-50

Schwarzschild, Bertram " Three Newly Discovered Exoplanets Have Masses Comparable to Neptune's", *Physics Today*, November 2004, pp. 27-29;

Sonett, C. P. and Chan, M. A. "Neoproterozoic Earth-Moon Dynamics: rework of 900 million ago Big Cottonwood Canyon tidal laminae" *Geophysics Research Letters*, 25(4), 539-542.

Sharma, B. K. "Theoretical Formulation of Earth-Moon System revisited," *Proceedings of Indian Science Congress 82nd Session*, 3rd January 1995 to 8th January 1995, Jadavpur University, Calcutta pp. 17.

Sharma, B. K. & Ishwar, B. "Planetary Satellite Dyanics : Earth-Moon, Mars-Phobos-Deimos and Pluto-Charon (Parth-I)" *35th COSPAR Scientific Assembly*, 18-25th July 2004, Paris, France

Sharma, B. K. & Ishwar, B. "A New Perspective on the Birth and Evolution of our Solar System based on Planetary Satellite Dynamics", *35th COSPAR Scientific Assembly*, 18-25th July 2004, Paris, France.

Sharma, B.K., Rangesh Neelmani and Ishwar, Bhola, "The software simulation of the spiral trajectory of our Moon",Advances in Space Research, Vol 23 (2009), 460-466;

Sharma,B.K."Architectural Design Rules of the birth and evolution of Solar Systems", Earth, Moon and Planets, Vol.108, Issue 1 (2011), 15-37;

Sharma,B.K., " The emergence of Primary-centric World View and its validation by Celebrated Hulse-Taylor Pulsar Binary and Pre-cataclysmic Binary NN Serpentis", submitted to *Earth,Moon and Planets*, MOON-S-11-00061.fdf (2011)

Sharma,B.K., "Primary-centric World-View proposes a Fractal Architecture of the Universe- A Post Copernican Conjecture", *Submission of full paper to Space Dynamics subsection of Mathematical Section of 99th Indian Science Congress- 3rd to 7th January (2012)*, at KIIT, Bhuneshwar, Orissa, India. Submitted to Annual Review of Astronomical and Astrophysics.(2012A)

Sharma,B.K., "Iapetus hypothetical sub-satellite re-visited and it reveals celestial body formation process criteria in the Primary-centric Framework", submitted to *42nd Scientific Assembly of COSPAR-2012*, Infosys Campus, Mysore, 16th-20th July 2012.(2012B)

Sharma, B.K., "Primary-centric World-View unravels the underlying Architecture of the Enigmatic Circumbinary Planet Kepler 16b System, Kepler 34b and Kepler 35b." Submitted to *NATURE* (2012C)

Sharma, B.K. "The probable evolutionary history of PSR J1719-1438 and its compacted companion", in preparation (2012D);

Singer, Dorothea Waley, " *Giordano Bruno, his Life and Thought*", New York Schuman, (1950)

Shiga, David " Imaging Exoplanets", *Sky & Telescope,* April 2004, pp.45-52;

Stillman,Drake, "Galileo at Work", Courier Dover Publications, pp.153.

Taylor, S. Rose and Mclennan Scott M. "The Evolution of Continental Crust," *Scientific American*, January, 1996.

Thommes, E.W. ;Matsumura, S.; Rasio,F.A. " Gas disks to gas giants: simulating the birth of planetary systems". *Science* 321, 814–817 (2008)

Tsiganis, K., Gomes, R., Morbidelli, A. and Levison, H.F. " Origin of the orbital architecture of the giant planets of our Solar System", *Nature*, Vol. 435, 26 May 2005, pp. 459-461;

Walker, J. C. G. & Zahnle, K. J. "Lunar Nodal Tide and distance to the Moon during Precambrian," *Nature*, 320, 600-602, (1986).

Wells, John W., *Nature*, 197, 948-950, 1963.

Wells, John W., *"Paleontological Evidence of the Rate of the Earth's Rotation"*, Earth-Moon System, edited by Marsden & Cameron, Plenum Press, 1966. pp. 70-81.

Welsh, W.F.; Orosze, J.A.; Carter, J.A. et.al. "Transiting Circumbinary Planets Kepler-34b and Kepler-35b", Nature 481,475-479, (26 Jan.2012) 10768/Letter doi:10.1038, (2012)

Williams, George E., "Geological Constraints on the Precambrian Hisotry of Earth's Rotation and the Moon's Orbit", *Review of Geophysics*, 38, 37-59. 1/February 2000.

Windleband, W. " Renessaince, Enlightment and Modern", *A History of Philosophy*, 2, Harper & Brothers, New York (1958).

Yates, Frances " *Giordano Bruno and the Hermetic Tradition"*, University of Chicago Press, (1964)

Yin, X. Zhour, Y., Pan, J., Zheng, D., Fang, M., Liao, X., He, M-X., Liu. W. T. and Ding, X. " Pacific warm pool excitation, earth rotation and El nino Southern

Oscillations",*Geophysical Research Letters*, Vol. 29, No 21, 2031, doi.10.1029/2002GLO15685 pp. 27-1 to 27-4, 2002.

Zeik/Gauntand editors *Astronomy* (IInd Edition), Cosmic Perspective,

Zimmerman, Robert " Exo-Earths", *Astronomy*, August 2004, pp.42-47;

Part 4

Cosmology and CMB Physics

7

Nonlinear Electrodynamics Effects on the Cosmic Microwave Background: Circular Polarization

Herman J. Mosquera Cuesta[1,2,3,4] and Gaetano Lambiase[5,6]
[1]Departamento de Física, Centro de Ciências Exatas e Tecnológicas (CCET),
Universidade Estadual Vale do Acaraú, Sobral, Ceará,
[2]Instituto de Cosmologia, Relatividade e Astrofísica (ICRA-BR),
Centro Brasileiro de Pesquisas Físicas, Urca Rio de Janeiro, RJ,
[3]International Center for Relativistic Astrophysics Network (ICRANet), Pescara,
[4]International Institute for Theoretical Physics and High Mathematics
Einstein-Galilei, PRATO,
[5]Dipartimento di Fisica "E. R. Caianiello",
Universitá di Salerno, Fisciano (Sa),
[6]INFN, Sezione di Napoli,
[1,2]Brazil
[3,4,5,6]Italy

1. Introduction

Historically, the modifications to standard electrodynamics were introduced for preventing the appearance of infinite physical quantities in theoretical analysis involving electromagnetic interactions. For instance, Born-Infeld [1] proposed a model in which the infinite self energy of point particles (typical of linear electrodynamics) are removed by introducing an upper limit on the electric field strength and by considering the electron an electrically charged particle of finite radius. Along this line, other Lagrangians for nonlinear electrodynamics were proposed by Plebanski, who also showed that Born-Infeld model satisfy physically acceptable requirements [20], due to its feature of being inspired on the special relativity principles. Applications and consequences of nonlinear electrodynamics have been extensively studied in literature, ranging from cosmological and astrophysical models [22] to nonlinear optics, high power laser technology and plasma physics [25].

In this paper we investigate the polarization of CMB photons if electrodynamics is inherently nonlinear. We compute the polarization angle of CMB photons propagating in an expanding Universe, by considering in particular cosmological models with planar symmetry. It is shown that the polarization does depend on the parameter characterizing the nonlinearity of electrodynamics, which is thus constrained by making use of the recent data from WMAP and BOOMERanG (for other models see [26]). It is worth to point out that the effect we are investigating, i.e. the rotation of the polarization angle as radiation propagates in a planar geometry, is strictly related to Skrotskii effect [27]. The latter is analogous to Faraday effect obtained for radiation propagating in a magnetic field. The effect on the CMB radiation as predicted by NLED is polarization *circular* in nature. This is a unique and very distinctive feature which can be falsified with the upcoming results to be released by the PLANCK satellite collaboration.

2. Some Lagrangian formulations of nonlinear electrodynamics

To start with, it is worth to recall that according to quantum electrodynamics (QED: see [7, 8] for a complete review on NLED and QED) a vacuum has nonlinear properties (Heisenberg & Euler 1936; Schwinger 1951) which affect the photon propagation. A noticeable advance in the realization of this theoretical prediction has been provided by [Burke, Field, Horton-Smith , etal., 1997), who demonstrated experimentally that the inelastic scattering of laser photons by gamma-rays in a background magnetic ield is definitely a nonlinear phenomenon. The propagation of photons in NLED has been examined by several authors [Bialynicka-Birula & Bialynicki-Birula, 1970; Garcia & Plebanski, 1989; Dittrich & Gies, 1998; De Lorenci, Klippert, Novello, etal., 2000; Denisov, Denisova & Svertilov, 2001a, 2001b, Denisov & Svertilov, 2003]. In the geometric optics approximation, it was shown by [Novello, De Lorenci, Salim & etal., 2000; Novello & Salim, 2001], that when the photon propagation is identified with the propagation of discontinuities of the EM field in a nonlinear regime, a remarkable feature appears: The discontinuities propagate along null geodesics of an *effective* geometry which depends on the EM field on the background. This means that the NLED interaction can be geometrized. An immediate consequence of this NLED property is the prediction of the phenomenon dubbed as photon acceleration, which is nothing else than a shift in the frequency of any photon traveling over background electromagnetic fields. The consequences of this formalism are examined next.

2.1 Heisenberg-Euler approach

The Heisenberg-Euler Lagrangian for nonlinear electrodynamics (up to order 2 in the truncated infinite series of terms involving F) has the form [16]

$$L_{H-E} = -\frac{1}{4}F + \bar{\alpha}F^2 + \bar{\beta}G^2,$$ (1)

where $F = F_{\mu\nu}F^{\mu\nu}$, with $F_{\mu\nu} = \partial_\mu A_\nu - \partial_\nu A_\mu$, and $G = \frac{1}{2}\eta_{\alpha\beta\gamma\delta}F^{\alpha\beta}F^{\gamma\delta} = -4\vec{E}\cdot\vec{B}$, with greek index running (0, 1, 2, 3), while $\bar{\alpha}$ and $\bar{\beta}$ are arbitrary constants.

When this Lagrangian is used to describe the photon dynamics the equations for the EM field in vacuum coincide in their form with the equations for a continuum medium in which the electric permittivity and magnetic permeability tensors $\epsilon_{\alpha\beta}$ and $\mu_{\alpha\beta}$ are functions of the electric and magnetic fields determined by some observer represented by its 4-vector velocity V^μ [Denisov, Denisova & Svertilov, 2001a, 2001b; Denisov & Svertilov, 2003; Mosquera Cuesta & Salim, 2004a, 2004b]. The attentive reader must notice that this first order approximation is valid only for B-fields smaller than $B_q = \frac{m^2c^3}{e\hbar} = 4.41 \times 10^{13}$ G (Schwinger's critical B-field [1]). In curved spacetime, these equations are written as

$$D^\alpha_{||\alpha} = 0, \qquad B^\alpha_{||\alpha} = 0,$$ (2)

$$D^\alpha_{||\beta}\frac{V^\beta}{c} + \eta^{\alpha\beta\rho\sigma}V_\rho H_{\sigma||\beta} = 0,$$ (3)

$$B^\alpha_{\|\beta} \frac{V^\beta}{c} - \eta^{\alpha\beta\rho\sigma} V_\rho E_{\sigma\|\beta} = 0 . \tag{4}$$

Here, the vertical bars subscript "$\|$" stands for covariant derivative and $\eta^{\alpha\beta\rho\sigma}$ is the antisymmetric Levi-Civita tensor.

The 4-vectors representing the electric and magnetic fields are defined as usual in terms of the electric and magnetic fields tensor $F_{\mu\nu}$ and polarization tensor $P_{\mu\nu}$

$$E_\mu = F_{\mu\nu} \frac{V^\nu}{c} , \qquad B_\mu = F^*_{\mu\nu} \frac{V^\nu}{c} , \tag{5}$$

$$D_\mu = P_{\mu\nu} \frac{V^\nu}{c} , \qquad H_\mu = P^*_{\mu\nu} \frac{V^\nu}{c} , \tag{6}$$

where the dual tensor $X^*_{\mu\nu}$ is defined as $X^*_{\mu\nu} = \frac{1}{2}\eta_{\mu\nu\alpha\beta}X^{\alpha\beta}$, for any antisymmetric second-order tensor $X_{\alpha\beta}$.

The meaning of the vectors D^μ and H^μ comes from the Lagrangian of the EM field, and in the vacuum case they are given by

$$H_\mu = \mu_{\mu\nu}B^\nu , \qquad D_\mu = \epsilon_{\mu\nu}E^\nu , \tag{7}$$

where the permeability and tensors are given as

$$\mu_{\mu\nu} = \left[1 + \frac{2\alpha}{45\pi B_q^2}\left(B^2 - E^2\right)\right]h_{\mu\nu} - \frac{7\alpha}{45\pi B_q^2}E_\mu E_\nu , \tag{8}$$

$$\epsilon_{\mu\nu} = \left[1 + \frac{2\alpha}{45\pi B_q^2}\left(B^2 - E^2\right)\right]h_{\mu\nu} + \frac{7\alpha}{45\pi B_q^2}B_\mu B_\nu . \tag{9}$$

In these expressions α is the EM coupling constant ($\alpha = \frac{e^2}{\hbar c} = \frac{1}{137}$). The tensor $h_{\mu\nu}$ is the metric induced in the reference frame perpendicular to the observers determined by the vector field V^μ.

Meanwhile, as we are assuming that $E^\alpha = 0$, then one gets

$$\epsilon^\alpha_\beta = \epsilon h^\alpha_\beta + \frac{7\alpha}{45\pi B_q^2}B^\alpha B_\beta \tag{10}$$

and $\mu_{\alpha\beta} = \mu h_{\alpha\beta}$. The scalars ϵ and μ can be read directly from Eqs.(8, 9) as

$$\epsilon \equiv \mu = 1 + \frac{2\alpha}{45\pi B_q^2}B^2 . \tag{11}$$

Applying condition (8) and the method in ([14]) to the field equations when $E^\alpha = 0$, we obtain the constraints $e^\mu\epsilon_{\mu\nu}k^\nu = 0$ and $b^\mu k_\mu = 0$ and the following equations for the discontinuity fields e_α and b_α:

$$\epsilon^{\lambda\gamma}e_\gamma k_\alpha \frac{V^\alpha}{c} + \eta^{\lambda\mu\rho\nu}\frac{V_\rho}{c}\left(\mu b_\nu k_\mu - \mu'\lambda_\alpha B_\nu k_\mu\right) = 0 , \tag{12}$$

$$b^\lambda k_\alpha \frac{V^\alpha}{c} - \eta^{\lambda\mu\rho\nu}\frac{V_\rho}{c}\left(e_\nu k_\mu\right) = 0 . \tag{13}$$

Isolating the discontinuity field from (12), substituting in equation (13), and expressing the products of the completely anti-symmetric tensors $\eta_{\nu\xi\gamma\beta}\eta^{\lambda\alpha\rho\mu}$ in terms of delta functions, we obtain

$$b^\lambda(k_\alpha k^\alpha)^2 + \left(\frac{\mu'}{\mu}l_\beta b^\beta k_\alpha B^\alpha + \frac{\beta B_\beta b^\beta B_\alpha k^\alpha}{\mu - \beta B^2}\right)k^\lambda +$$

$$\left(\frac{\mu'}{\mu l_\alpha b^\alpha}(k_\beta V^\beta)^2(k_\alpha k^\alpha)^2 - \frac{\beta B_\alpha b^\alpha(k_\beta k^\beta)^2}{\mu - \beta B^2}\right)B^\lambda - \left(\frac{\mu'}{\mu}l_\mu b^\mu k_\alpha B^\alpha k_\beta V^\beta\right)V^\lambda = 0. \quad (14)$$

This expression is already squared in k_μ but still has an unknown b_α term. To get rid of it, one multiplies by B_λ, to take advantage of the EM wave polarization dependence. By noting that if $B^\alpha b_\alpha = 0$ one obtains the *dispersion relation* by separating out the $k^\mu k^\nu$ term, what remains is the (-) effective metric. Similarly, if $B_\alpha b^\alpha \neq 0$, one simply divides by $B_\gamma b^\gamma$ so that by factoring out $k^\mu k^\nu$, what results is the (+) effective metric. For the case $B_\alpha b^\alpha = 0$, one obtains the standard dispersion relation

$$g^{\alpha\beta}k_\alpha k_\beta = 0. \quad (15)$$

whereas for the case $B_\alpha b^\alpha \neq 0$, the result is

$$\left[\left(1 + \frac{\mu'B}{\mu} + \frac{\tilde{\beta}B^2}{\mu - \tilde{\beta}B^2}\right)g^{\alpha\beta} - \frac{\mu'B}{\mu}\frac{V^\alpha V^\beta}{c^2} + \left(\frac{\mu'B}{\mu} + \frac{\tilde{\beta}B^2}{\mu - \tilde{\beta}B^2}\right)l^\alpha l^\beta\right]k_\alpha k_\beta = 0, \quad (16)$$

where (') stands for $\frac{d}{dB}$, and we have defined

$$\tilde{\beta} = \frac{7\alpha}{45\pi B_q^2}, \quad \text{and} \quad l^\mu \equiv \frac{B^\mu}{|B^\gamma B_\gamma|^{1/2}} \quad (17)$$

as the unit 4-vector along the B-field direction.

From the above expressions we can read the effective metric $g_+^{\alpha\beta}$ and $g_-^{\alpha\beta}$, where the labels "+" and "-" refers to extraordinary and ordinary polarized rays, respectively. Then, we need the covariant form of the metric tensor, which is obtained from the expression defining the inverse metric $g_{\mu\nu}g^{\nu\alpha} = \delta_\mu^\alpha$. So that one gets from one side

$$g_{\mu\nu}^- = g_{\mu\nu} \quad (18)$$

and from the other

$$g_{\mu\nu}^+ = \left(1 + \frac{\mu'B}{\mu} + \frac{\beta B^2}{\mu - \beta B^2}\right)^{-1}g_{\mu\nu}$$

$$+ \left[\frac{\mu'B}{\mu(1 + \frac{\mu'B}{\mu} + \frac{\beta B^2}{\mu - \beta B^2})(1 + \frac{\beta B^2}{\mu - \beta B^2})}\right]\frac{V_\mu V_\nu}{c^2} + \left(\frac{\frac{\mu'B}{\mu} + \frac{\beta B^2}{\mu - \beta B^2}}{1 + \frac{\mu'B}{\mu} + \frac{\beta B^2}{\mu - \beta B^2}}\right)l_\mu l_\nu. \quad (19)$$

The function $\frac{\mu'B}{\mu}$ can be expressed in terms of the magnetic permeability of the vacuum, and is given as

$$\frac{\mu'B}{\mu} = 2\left(1 - \frac{1}{\mu}\right) .$$ (20)

Thus equation (19) indicates that the photon propagates on an effective metric.

2.2 Born-Infeld theory

The propagation of light can also be viewed within the framework of the Born-Infeld Lagrangian. Such theory is inspired in the special theory of relativity, and indeed it incorporates the principle of relativity in its construction, since the fact that nothing can travel faster than light in a vacuum is used as a guide to establishing the existence of an upper limit for the strength of electric fields around an isolated charge, an electron for instance. Such charge is then forced to have a characteristic size [5]. The Lagrangian then reads

$$L = -\frac{b^2}{2}\left[\left(1 + \frac{F}{b^2}\right)^{1/2} - 1\right] .$$ (21)

As in this particular case, the Lagrangian is a functional of the invariant F, i.e., $L = L(F)$, but not of the invariant $G \equiv B_\mu E^\mu$, the study of the NLED effects turns out to be simpler (here again we suppose $E = 0$). In the equation above, $b = \frac{e}{R_0^2} = \frac{e}{\frac{e^4}{m_0^2 c^8}} = \frac{m_0^2 c^8}{e^3} = 9.8 \times 10^{15}$ e.s.u.

In order to derive the effective metric that can be deduced from the B-I Lagrangian, one has therefore to work out, as suggested in the Appendix, the derivatives of the Lagrangian with respect to the invariant F. The first and second derivatives then reads

$$L_F = \frac{-1}{4\left(1 + \frac{F}{b^2}\right)^{1/2}} \quad \text{and} \quad L_{FF} = \frac{1}{8b^2\left(1 + \frac{F}{b^2}\right)^{3/2}} .$$ (22)

The $L(F)$ B-I Lagrangian produces an *effective* contravariant metric given as

$$g_{\text{eff}}^{\mu\nu} = \frac{-1}{4\left(1 + \frac{F}{b^2}\right)^{1/2}}g^{\mu\nu} + \frac{B^2}{2b^2\left(1 + \frac{F}{b^2}\right)^{3/2}}[h^{\mu\nu} + l^\mu l^\nu] .$$ (23)

Both the tensor $h_{\mu\nu}$ and the vector l^μ in this equation were defined earlier (see Eqs.(9) and (16) above).

Because the geodesic equation of the discontinuity (that defines the effective metric, see the Appendix) is conformal invariant, one can multiply this last equation by the conformal factor $-4\left(1 + \frac{F}{b^2}\right)^{3/2}$ to obtain

$$g_{\text{eff}}^{\mu\nu} = \left(1 + \frac{F}{b^2}\right)g^{\mu\nu} - \frac{2B^2}{b^2}[h^{\mu\nu} + l^\mu l^\nu] .$$ (24)

Then, by noting that

$$F = F_{\mu\nu}F^{\mu\nu} = -2(E^2 - B^2) ,$$ (25)

and recalling our assumption $E = 0$, then one obtains $F = 2B^2$. Therefore, the effective metric reads

$$g_{\text{eff}}^{\mu\nu} = \left(1 + \frac{2B^2}{b^2}\right) g^{\mu\nu} - \frac{2B^2}{b^2} \left[h^{\mu\nu} + l^{\mu} l^{\nu}\right] , \tag{26}$$

or equivalently

$$g_{\text{eff}}^{\mu\nu} = g^{\mu\nu} + \frac{2B^2}{b^2} V^{\mu} V^{\nu} - \frac{2B^2}{b^2} l^{\mu} l^{\nu} . \tag{27}$$

As one can check, this effective metric is a functional of the background metric $g^{\mu\nu}$, the 4-vector velocity field of the inertial observers V^{ν}, and the spatial configuration (orientation l^{μ}) and strength of the B-field.

Thus the covariant form of the background metric can be obtained by computing the inverse of the effective metric $g_{\text{eff}}^{\mu\nu}$ just derived. With the definition of the inverse metric $g_{\text{eff}}^{\mu\nu} g_{\nu\alpha}^{\text{eff}} = \delta^{\mu}{}_{\alpha}$, the covariant form of the effective metric then reads

$$g_{\mu\nu}^{\text{eff}} = g_{\mu\nu} - \frac{2B^2/b^2}{(2B^2/b^2 + 1)} V_{\mu} V_{\nu} + \frac{2B^2/b^2}{(2B^2/b^2 + 1)} l_{\mu} l_{\nu} , \tag{28}$$

which is the result that we were looking for. The terms additional to the background metric $g_{\mu\nu}$ characterize any effective metric.

2.3 Pagels-Tomboulis Abelian theory

In 1978, the Pagels-Tomboulis nonlinear Lagrangian for electrodynamics appeared as an effective model of an Abelian theory introduced to describe a perturbative gluodynamics model. It was intended to investigate the non trivial aspects of quantum-chromodynamics (QCD) like the asymptotic freedom and quark confinement [28]. In fact, Pagels and Tomboulis argued that:

"since in asymptotically free Yang-Mills theories the quantum ground state is not controlled by perturbation theory, there is no a priori reason to believe that individual orbits corresponding to minima of the classical action dominate the Euclidean functional integral. "

In view of this drawback, of the at the time understanding of ground states in quantum theory, they decided to examine and classify the vacua of the quantum gauge theory. To this goal, they introduced an effective action in which the gauge field coupling constant g is replaced by the effective coupling $\bar{g}(t) \cdot T = \ln \left[\frac{F_{\mu\nu}^a F^{a\,\mu\nu}}{\mu^4}\right]$. The vacua of this model correspond to paramagnetism and perfect paramagnetism, for which the gauge field is $F_{\mu\nu}^a = 0$, and ferromagnetism, for which $F_{\mu\nu}^a F^{a\,\mu\nu} = \lambda^2$, which implies the occurrence of spontaneous magnetization of the vacuum. [1] They also found no evidence for instanton solutions to the quantum effective action. They solved the equations for a point classical source of color spin, which indicates that in the limit of spontaneous magnetization the infrared energy of the field becomes linearly divergent. This leads to bag formation, and to an electric Meissner effect confining the bag contents.

This effective model for the low energy (3+1) QCD reduces, in the Abelian sector, to a nonlinear theory of electrodynamics whose density Lagrangian $L(X, Y)$ is a functional of the

---

[1] This is the imprint that such theory describes nonlinear electrodynamics.

invariants $X = F_{\mu\nu}F^{\mu\nu}$ and their dual $Y = (F_{\mu\nu}F^{\mu\nu})^{\star}$, having their equations of motion given by

$$\nabla_\mu \left(-L_X F^{\mu\nu} - L_Y{}^* F^{\mu\nu}\right) = 0, \tag{29}$$

where $L_X = \partial L/\partial X$ and $L_Y = \partial L/\partial Y$. This equation is supplemented by the Faraday equation, i. e., the electromagnetic field tensor cyclic identity (which remains unchanged)

$$\nabla_\mu F_{\nu\lambda} + \nabla_\nu F_{\lambda\mu} + \nabla_\lambda F_{\mu\nu} = 0. \tag{30}$$

In the case of a simple dependence on X, the equations of motion turn out to be [24] (here we put $C = 0$ and $4\gamma = -(\Lambda^8)^{(\delta-1)/2}$ in the original Lagrangian given in [28])

$$L_\delta = -\frac{1}{4}\left(\frac{X^2}{\Lambda^8}\right)^{(\delta-1)/2} X, \tag{31}$$

where δ is an dimensionless parameter and $[\Lambda] = (an\, energy\, scale)$. The value $\delta = 1$ yields the standard Maxwell electrodynamics.

The energy-momentum tensor for this Lagrangian $L(X)$ can be computed by following the standard recipe, which then gives

$$T_{\mu\nu} = \frac{1}{4\pi}\left(L_X g^{ab} F_{\mu a} F_{b\nu} + g_{\mu\nu} L\right) \tag{32}$$

while its trace turns out to be

$$T = -\frac{1-\delta}{\pi}\left(\frac{X^2}{\Lambda^8}\right)^{(\delta-1)/2} X. \tag{33}$$

It can be shown [24] that the positivity of the $T_0^0 \equiv \rho$ component implies that $\delta \geq 1/2$.

The Lagrangian (31) has been studied by [24] for explaining the amplification of the primordial magnetic field in the Universe, being the analysis focused on three different regimes: 1) $B^2 \gg E^2$, 2) $B^2 \simeq \mathcal{O}(E^2)$, 3) $E^2 \ll B^2$. It has also been used by [23] to discuss both the origin of the baryon asymmetry in the universe and the origin of primordial magnetic fields. More recently it has also been discussed in the review on " Primordial magneto-genesis" by [17].

Because the equation of motion (29) above, exhibits similar mathematical aspect as eq. (35) (reproduced in the Section), it appears clear that the Pagels and Tomboulis Lagrangian (31) leads also to an effective metric identical to that one given in equation (40), below.

2.4 Novello-Pérez Bergliaffa-Salim NLED

More recently, [21] Novello, Pérez Bergliaffa, Salim revisited the several general properties of nonlinear electrodynamics by assuming that the action for the electromagnetic field is that of Maxwell with an extra term, namely[2]

$$S - \int \sqrt{-g}\left(-\frac{F}{4} + \frac{\gamma}{F}\right) d^4 x, \tag{34}$$

---

[2] Notice that this Lagrangian is gauge invariant, and that hence charge conservation is guaranteed in this theory.

where $F \equiv F_{\mu\nu}F^{\mu\nu}$.

Physical motivations for bringing in this theory have been provided in [21]. Besides of those arguments, an equally unavoidable motivation comes from the introduction in the 1920's of both the Heisenberg-Euler and Born-Infeld nonlinear electrodynamics discussed above, which are valid in the regime of extremely high magnetic field strengths, i.e. near the Schwinger's limit. Both theories have been extensively investigated in the literature (see for instance [22] and the long list of references therein). Since in nature non only such very strong magnetic fields exist, then it appears to be promising to investigate also those super weak field frontiers. From the conceptual point of view, this phenomenological action has the advantage that it involves only the electromagnetic field, and does not invoke entities that have not been observed (like scalar fields) and/or speculative ideas (like higher-dimensions and brane worlds).

At first, one notices that for high values of the field F, the dynamics resembles Maxwell's one except for small corrections associate to the parameter γ, while at low strengths of F it is the $1/F$ term that dominates. (Clearly, this term should dramatically affect, for instance, the photon-$\vec{B}$ field interaction in intergalactic space, which is relevant to understand the solution to the Pioneer anomaly using NLED.). The consistency of this theory with observations, including the recovery of the well-stablished Coulomb law, was shown in [21] using the cosmic microwave radiation bound, and also after discussing the anomaly in the dynamics of Pioneer 10 spacecraft [22]. Both analysis provide small enough values for the coupling constant γ.

2.4.1 Photon dynamics in NPS NLED: Effective geometry

Next we investigate the effects of nonlinearities in the evolution of EM waves in the vacuum permeated by background $\vec{B}$-fields. An EM wave is described onwards as the surface of discontinuity of the EM field. Extremizing the Lagrangian $L(F)$, with $F(A_\mu)$, with respect to the potentials A_μ yields the following field equation [20]

$$\nabla_\nu(L_F F^{\mu\nu}) = 0, \tag{35}$$

where ∇_ν defines the covariant derivative. Besides this, we have the EM field cyclic identity

$$\nabla_\nu F^{*\mu\nu} = 0 \quad \Leftrightarrow \quad F_{\mu\nu|\alpha} + F_{\alpha\mu|\nu} + F_{\nu\alpha|\mu} = 0. \tag{36}$$

Taking the discontinuities of the field Eq.(35) one gets (all the definitions introduced here are given in [14]) [3]

$$L_F f_\lambda{}^\mu k^\lambda + 2L_{FF}F^{\alpha\beta}f_{\alpha\beta}F^{\mu\lambda}k_\lambda = 0, \tag{37}$$

which together with the discontinuity of the Bianchi identity yields

$$f_{\alpha\beta}k_\gamma + f_{\gamma\alpha}k_\beta + f_{\beta\gamma}k_\alpha = 0. \tag{38}$$

---

[3] Following Hadamard's method [15], the surface of discontinuity of the EM field is denoted by Σ. The field is continuous when crossing Σ, while its first derivative presents a finite discontinuity. These properties are specified as follows: $\left[F_{\mu\nu}\right]_\Sigma = 0$, $\left[F_{\mu\nu|\lambda}\right]_\Sigma = f_{\mu\nu}k_\lambda$, where the symbol $\left[F_{\mu\nu}\right]_\Sigma = \lim_{\delta\to 0^+}(J|_{\Sigma+\delta} - J|_{\Sigma-\delta})$ represents the discontinuity of the arbitrary function J through the surface Σ. The tensor $f_{\mu\nu}$ is called the discontinuity of the field, $k_\lambda = \partial_\lambda\Sigma$ is the propagation vector, and the symbols "$|$" and "$\|$" stand for partial and covariant derivatives.

A scalar relation can be obtained if we contract this equation with $k^\gamma F^{\alpha\beta}$, which yields

$$(F^{\alpha\beta} f_{\alpha\beta} g^{\mu\nu} + 2F^{\mu\lambda} f_\lambda{}^\nu) k_\mu k_\nu = 0 . \tag{39}$$

It is straightforward to see that here we find two distinct solutions: a) when $F^{\alpha\beta} f_{\alpha\beta} = 0$, case in which such mode propagates along standard null geodesics, and b) when $F^{\alpha\beta} f_{\alpha\beta} = \chi$. In the case a) it is important to notice that in the absence of charge currents, this discontinuity describe the propagation of the wave front as determined by the field equation (35), above. Thence, following [19] the quantity $f^{\alpha\beta}$ can be decomposed in terms of the propagation vector k_α and a space-like vector a_β (orthogonal to k_α) that describes the wave polarization. Thus, only the light-ray having polarization and direction of propagation such that $F^{\alpha\beta} k_\alpha a_\beta = 0$ will follow geodesics in $g_{\mu\nu}$. Any other light-ray will propagate on the effective metric (40). Meanwhile, in this last case, we obtain from equations (37) and (39) the propagation equation for the field discontinuities being given by [22]

$$\underbrace{\left(g^{\mu\nu} - 4\frac{L_{FF}}{L_F} F^{\mu\alpha} F_\alpha{}^\nu \right)}_{\text{effective metric}} k_\mu k_\nu = 0 . \tag{40}$$

This equation proves that photons propagate following a geodesic that is not that one on the background space-time, $g^{\mu\nu}$, but rather they follow the *effective metric* given by Eq.(40), which depends on the background field $F^{\mu\alpha}$, i. e., on the $\vec{B}$-field.

3. Minimally coupling gravity to nonlinear electrodynamics

The action of (nonlinear) electrodynamics coupled minimally to gravity is

$$S = \frac{1}{2\kappa} \int d^4x \sqrt{-g} R + \frac{1}{4\pi} \int d^4x \sqrt{-g} L(X, Y) , \tag{41}$$

where $\kappa = 8\pi G$, L is the Lagrangian of nonlinear electrodynamics depending on the invariant $X = \frac{1}{4} F_{\mu\nu} F^{\mu\nu} = -2(\mathbf{E}^2 - \mathbf{B}^2)$ and $Y = \frac{1}{4} F_{\mu\nu} {}^*F^{\mu\nu}$, where $F^{\mu\nu} \equiv \nabla_\mu A_\nu - \nabla_\nu A_\mu$, and $^*F^{\mu\nu} = \epsilon^{\mu\nu\rho\sigma} F_{\rho\sigma}$ is the dual bivector, and $\epsilon^{\alpha\beta\gamma\delta} = \frac{1}{2\sqrt{-g}} \varepsilon^{\alpha\beta\gamma\delta}$, with $\varepsilon^{\alpha\beta\gamma\delta}$ the Levi-Civita tensor ($\varepsilon_{0123} = +1$).

The equations of motion are

$$\nabla_\mu \left(-L_X F^{\mu\nu} - L_Y {}^*F^{\mu\nu} \right) = 0 , \tag{42}$$

where $L_X = \partial L / \partial X$ and $L_Y = \partial L / \partial Y$,

$$\nabla_\mu F_{\nu\lambda} + \nabla_\nu F_{\lambda\mu} + \nabla_\lambda F_{\mu\nu} = 0 . \tag{43}$$

After a swift grasp on this set of equations one realizes that is difficult to find solutions in closed form of these equations. Therefore to study the effects of nonlinear electrodynamics, we confine ourselves to consider the abelian Pagels-Tomboulis theory [28], proposed as an effective model of low energy QCD. The Lagrangian density of this theory involves only the invariant X in the form

$$L(X) = -\left(\frac{X^2}{\Lambda^8} \right)^{\frac{\delta-1}{2}} X = -\gamma X^\delta , \tag{44}$$

where γ (or Λ) and δ are free parameters that, with appropriate choice, reproduce the well known Lagrangian already studied in the literature. γ has dimensions [energy]$^{4(1-\delta)}$.

The equation of motion for the Pagels-Tomboulis theory follows from Eq. (42) with $Y = 0$

$$\nabla_\mu F^{\mu\nu} = -(\delta - 1)\frac{\nabla_\mu X}{X} F^{\mu\nu}. \tag{45}$$

In terms of the potential vector A^μ, and imposing the Lorentz gauge $\nabla_\mu A^\mu = 0$, Eq. (45) becomes

$$\nabla_\mu \nabla^\mu A^\nu + R^\nu_\mu A^\mu = -(\delta - 1)\frac{\nabla_\mu X}{X}(\nabla^\mu A^\nu - \nabla^\nu A^\mu), \tag{46}$$

where the Ricci tensor R^ν_μ appears because the relation $[\nabla^\mu, \nabla_\nu]A^\nu = -R^\mu_\mu A^\mu$.

We work in the geometrical optics approximation (this means that the scales of variation of the electromagnetic fields are smaller than the cosmological scales), so that the 4-vector $A^\mu(x)$ can be written as [29]

$$A^\mu(x) = Re\left[(a^\mu(x) + \epsilon b^\mu(x) + \ldots)e^{iS(x)/\epsilon}\right] \tag{47}$$

with $\epsilon \ll 1$ so that the phase S/ϵ varies faster than the amplitude. By defining the wave vector $k_\mu = \nabla_\mu S$, which defines the direction of the photon propagation, one finds that the gauge condition implies $k_\mu a^\mu = 0$ and $k_\mu b^\mu = 0$. It turns out to be convenient to introduce the normalized polarization vector ε^μ so that the vector a^μ can be written as

$$a^\mu(x) = A(x)\varepsilon^\mu, \qquad \varepsilon_\mu \varepsilon^\mu = 1. \tag{48}$$

As a consequence of (48), one also finds $k_\mu \varepsilon^\mu = 0$, i.e. the wave vector is orthogonal to the polarization vector.

To leading order in ϵ, the relevant equations are

$$(2\delta - 1)k^2 = 0 \quad \rightarrow \quad k_\mu k^\mu = 0, \tag{49}$$

for $\delta \neq \frac{1}{2}$, hence photons propagate along null geodesics, and

$$k^\mu \nabla_\mu \varepsilon^\sigma = \frac{\delta - 1}{\delta} k_\mu \left[\nabla^\sigma \varepsilon^\mu - (\varepsilon^\rho \nabla_\rho \varepsilon^\mu)\varepsilon^\sigma\right]. \tag{50}$$

4. Space-time anisotropy and magnetic energy density evolution

In what follows we consider cosmological models with planar symmetry

$$ds^2 = dt^2 - b^2(dx^2 + dy^2) - c^2 dz^2, \tag{51}$$

where $b(t)$ and $c(t)$ are the scale factors, which are normalized in order that $b(t_0) = 1 = c(t_0)$ at the present time t_0. As Eq. (51) shows, the symmetry is on the (xy)-plane.

We shall assume that photons propagate along the (positive) x-direction, so that $k^\mu = (k^0, k^1, 0, 0)$. Gauge invariance assures that the polarization vector of photons has only two independent components, which are orthogonal to the direction of the photons motion.

By defining the affine parameter λ which measures the distance along the line-element, $k^\mu \equiv dx^\mu/d\lambda$, one obtains that ε^2 and ε^3 satisfy the following geodesic equation (from Eq. (50))

$$\frac{1}{k^0}\mathcal{D}\ln(b\varepsilon^2) = -\frac{\delta-1}{\delta}\left(-\frac{\dot{b}}{b}+\frac{\dot{c}}{c}\right)(c\varepsilon^3)^2, \tag{52}$$

$$\frac{1}{k^0}\mathcal{D}\ln(c\varepsilon^3) = -\frac{\delta-1}{\delta}\left(-\frac{\dot{c}}{c}+\frac{\dot{b}}{b}\right)(b\varepsilon^2)^2. \tag{53}$$

where $\mathcal{D} \equiv k^\mu\nabla_\mu$. Moreover, the difference of the Hubble expansion rate $\dot{b}/b$ and $\dot{c}/c$ can be written as

$$\frac{\dot{b}}{b} - \frac{\dot{c}}{c} = \frac{1}{2(1-e^2)}\frac{de^2}{dt} \tag{54}$$

where we have introduced the eccentricity

$$e(t) = \sqrt{1-\left(\frac{c}{b}\right)^2}. \tag{55}$$

The polarization angle α is defined as $\alpha = \arctan[(c\varepsilon^3)/(b\varepsilon^2)]$. By introducing the reference time t, corresponding to the moment in which photons are emitted from the last scattering surface, and the instant t_0, corresponding to the present time, one finally gets

$$\Delta\alpha = \alpha(t) - \alpha(t_0) = \frac{\delta-1}{4\delta}Ke^2(z), \tag{56}$$

where K is a constant. Here we have used $e(t_0) = 0$, because of the normalization condition $b(t_0) = c(t_0) = 1$, and $\log(1-e^2) \sim -e^2$.

Notice that for $\delta = 1$ or $e^2 = 0$ there is no rotation of the polarization angle, as expected. Moreover, in the case in which photons propagate along the direction z-direction, so that $\bar{k}^a = (\omega_0,0,0,k)$, we find that the NLED have no effects as concerns to the rotation of the polarization angle.

4.1 Eccentricity evolution on cosmic time

The time evolution of the eccentricity is determined from the Einstein field equations

$$\frac{1}{1-e^2}\frac{d(e\dot{e})}{dt} + 3H_b(e\dot{e}) + \frac{(e\dot{e})^2}{(1-e^2)^2} = 2\kappa\rho_B, \tag{57}$$

where $H_b = \dot{b}/b$ and

$$\rho_B = \frac{B^2}{8\pi}\left(\frac{B^2}{2\Lambda^4}\right)^{\delta-1}. \tag{58}$$

It is extremely difficult to exactly solve this equation. We shall therefore assume that the e^2-terms can be neglected. Since $b(t) \sim t^{2/3}$ during the matter-dominated era, Eq. (57) implies

$$e^2(z) = 18F_\delta(z)\Omega_B^{(0)}, \tag{59}$$

where we used $1+z = b(t_0)/b(t)$, $e(t_0) = 0$, and

$$F_\delta \equiv \frac{3}{(9-8\delta)(4\delta-3)} - 2 - \frac{3(1+z)^{4\delta-3}}{(9-8\delta)(4\delta-3)} + 2(1+z)^{\frac{3}{2}}. \tag{60}$$

$\Omega_B^{(0)}$ is the present energy density ratio

$$\Omega_B^{(0)} = \frac{\rho_B}{\rho_{cr}} = \frac{B^2(t_0)}{8\pi\rho_{cr}} \left(\frac{B^2(t_0)}{2\Lambda^4}\right)^{\delta-1} \simeq \tag{61}$$

$$10^{-11} \left(\frac{B(t_0)}{10^{-9}\text{G}}\right)^2 \left(\frac{B^2(t_0)}{2\Lambda^4}\right)^{\delta-1},$$

with $\rho_{cr} = 3H_b^2(t_0)/\kappa = 8.1h^2 10^{-47}$ GeV^4 ($h = 0.72$ is the little-h constant), and $B(t_0)$ is the present magnetic field amplitude.

From Eq. (56) then follows

$$\Delta\alpha = \frac{\delta-1}{4\delta} K e^2(z_{dec}). \tag{62}$$

where $e(z_{dec})^2$ the eccentricity (59) evaluated at the decoupling $z = 1100$.

4.2 Constraints on extragalactic B strengths

To make an estimate on the parameter δ, we need the order of amplitude of the present magnetic field strength $B(t_0)$. In this respect, observations indicate that there exist, in cluster of galaxies, magnetic fields with field strength $(10^{-7} - 10^{-6})$ G on 10 kpc - 1 Mpc scales, whereas in galaxies of all types and at cosmological distances, the order of magnitude of the magnetic field strength is $\sim 10^{-6}$ G on (1-10) kpc scales. The present accepted estimations is (see for example [30])

$$B(t_0) \lesssim 10^{-9} \text{ G}. \tag{63}$$

Moreover, for an ellipsoidal Universe the eccentricity satisfies the relation $0 \le e^2 < 1$. The condition $e^2 > 0$ means $F_\delta > 0$, with F_δ defined in (60). The function F_δ given by Eq. (60) is represented in Fig. 1. Clearly the allowed region where F_δ is positive does depend on the redshift z. On the other hand, the condition $e^2 < 1$ poses constraints on the magnetic field strength. By requiring $e^2 < 10^{-1}$ (in order that our approximation to neglect e^2-terms in (57) holds), from Eqs. (59)-(61) it follows

$$B(t_0) \lesssim 9 \times 10^{-8}\text{G}. \tag{64}$$

It must also be noted that such magnetic fields does not affect the expansion rate of the universe and the CMB fluctuations because the corresponding energy density is negligible with respect to the energy density of CMB.

5. Stokes parameters, rotated CMB spectra and constraints on parameter Λ

The propagation of photons can be described in terms of the Stokes parameters I, Q, U, and V. The parameters Q and V can be decomposed in gradient-like (G) and a curl-like (C) components [31] (G and C are also indicated in literature as E and B), and characterize the orthogonal modes of the linear polarization (they depend on the axes where the linear polarization are defined, contrarily to the physical observable I and V which are independent on the choice of coordinate system).

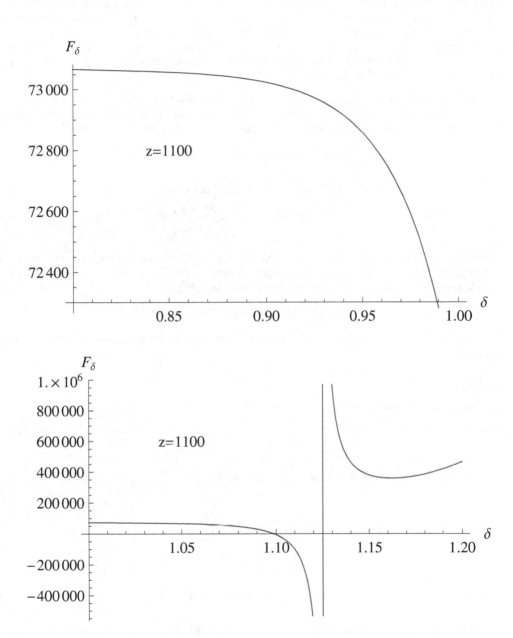

Fig. 1. In this plot is represented F_δ vs δ for $\delta \leq 1$ (upper plot) and $\delta \geq 1$ (lower plot). The condition that the eccentricity is positive follows for $F_\delta > 0$.

The polarization G and C and the temperature (T) are crucial because they allow to completely characterize the CMB on the sky. If the Universe is isotropic and homogeneous and the electrodynamics is the standard one, then the TC and GC cross-correlations power spectrum vanish owing to the absence of the cosmological birefringence. In presence of the latter, on the contrary, the polarization vector of each photons turns out to be rotated by the angle $\Delta\alpha$, giving rise to TC and GC correlations.

Using the expression for the power spectra $C_l^{XY} \sim \int dk[k^2 \Delta_X(t_0)\Delta_Y(t_0)]$, where $X, Y = T, G, C$ and Δ_X are the polarization perturbations whose time evolution is controlled by the Boltzman equation, one can derive the correlation for T, G and C in terms of $\Delta\alpha$ [32]

$$C_l'^{TC} = C_l^{TC} \sin 2\Delta\alpha, \quad C_l'^{TG} = C_l^{TG} \cos 2\Delta\alpha, \tag{65}$$

$$C_l'^{GC} = \frac{1}{2}\left(C_l^{GG} - C_l^{CC}\right)\sin 4\Delta\alpha, \tag{66}$$

$$C_l'^{GG} = C_l^{GG} \cos^2 2\Delta\alpha + C_l^{CC}\sin^2 2\Delta\alpha, \tag{67}$$

$$C_l'^{CC} = C_l^{CC} \cos^2 2\Delta\alpha + C_l^{GG}\sin^2 2\Delta\alpha. \tag{68}$$

The prime indicates the rotated quantities. Notice that the CMB temperature power spectrum remains unchanged under the rotation. Experimental constraints on $\Delta\alpha$ have been put from the observation of CMB polarization by WMAP and BOOMERanG

$$\Delta\alpha = (-2.4 \pm 1.9)^\circ = [-0.0027\pi, -0.0238\pi]. \tag{69}$$

The combination of Eqs. (69) and (62), and the laboratory constraints $|\delta - 1| \ll 1$ allow to estimate Λ.

5.1 Estimative of Λ

To estimate Λ we shall write

$$B = 10^{-9+b}\, G \qquad b \lesssim 2, \tag{70}$$

$$F_\delta = 2z^{3/2} \qquad z = 1100 \gg 1. \tag{71}$$

The bound (69) can be therefore rewritten in the form

$$\frac{10^{-3}}{\mathcal{A}} \lesssim |\delta - 1| \lesssim \frac{10^{-2}}{\mathcal{A}}, \tag{72}$$

where

$$\mathcal{A} \equiv \frac{9K}{14} F_\delta \Omega_B^{(0)} \simeq \tag{73}$$

$$\simeq K \times 10^{-6+2b} \left[0.24 \times 10^{-56+2b}\left(\frac{\text{GeV}}{\Lambda}\right)^4\right]^{\delta-1}.$$

The condition $|\delta - 1| \ll 1$ requires $\mathcal{A} \gg 1$. Setting $\mathcal{A} = 10^a$, with $a > \mathcal{O}(1)$, it follows

$$\text{Log}\left[\frac{\Lambda}{\text{GeV}}\right] = \left(-14 + \frac{b}{2}\right) - \frac{a - 2b + 6 - \text{Log}K}{4(\delta - 1)}. \tag{74}$$

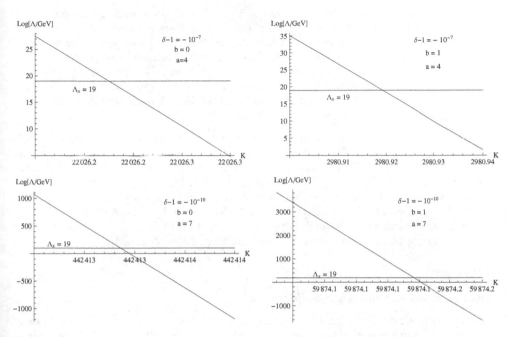

Fig. 2. Λ vs K for different values of the parameter $\delta - 1$, a and b. The parameter a is related to the range in which $\delta - 1$ varies, i.e. $-10^{-3-a} \lesssim \delta - 1 \lesssim -10^{-2-a}$, while b parameterizes the magnetic field strength $B = 10^{-9+b}$ G. The red-shift is $z = 1100$. Plot refers to Planck scale $\Lambda = 10^{\Lambda_x}$ GeV, with $\Lambda_{x=Pl} = 19$. Similar plots can be also obtained for GUT ($\Lambda_{GUT} = 16$) and EW ($\Lambda_{EW} = 3$) scales.

The constant K can now be determined to fix the characteristic scale Λ. Writing $\Lambda = 10^{\Lambda_x}$ GeV, where $\Lambda_{x=Pl} = 19$, i.e. Λ is fixed by Planck scale, Eq. (74) yields

$$K = 10^{a-2b+6-\zeta},$$

$$\zeta \equiv \frac{4(\delta - 1)\Lambda_x}{14 - b/2} \ll 1. \tag{75}$$

In Fig. 2 is plotted $\text{Log}(\Lambda/\text{GeV})$ vs K for fixed values of the parameters a, b and $\delta - 1$. Similar plots can be derived for GUT and EW scales.

6. Discussion and future perspective

In conclusion, in this paper we have calculated, in the framework of the nonlinear electrodynamics, the rotation of the polarization angle of photons propagating in a Universe with planar symmetry. We have found that the rotation of the polarization angle does depend on the parameter δ, which characterizes the degree of nonlinearity of the electrodynamics. This parameter can be constrained by making use of recent data from WMAP and BOOMERang. Results show that the CMB polarization signature, if detected by future CMB observations, would be an important test in favor of models going beyond the standard model, including the nonlinear electrodynamics.

Moreover, an interesting topic that deserves to be studied is the possibility to generate a circular polarization in nonlinear electrodynamics. As well know there is at the moment no mechanism to generate it at the last scattering, provided some extension of the standard electrodynamics [33]. Recently, Motie and Xue [34] have discussed the photon circular polarization in the framework of Euler-Heisenberg Lagrangian. It is therefore worthwhile to derive a circular polarization in the framework of the Pagels-Tomboulis nonlinear electrodynamics described by (44) and for a background described by nonplanar geometry. It is expected, however, that the effects are very small [35].

As closing remark, we would like to point out that the approach to analyze the CMB polarization in the context of NLED that we have presented above can also be applied to discuss the extreme-scale alignments of quasar polarization vectors [36], a cosmic phenomenon that was discovered by Hutsemekers[37] in the late 1990's, who presented paramount evidence for very large-scale coherent orientations of quasar polarization vectors (see also Hutsemekers and Lamy [38]). As far as the authors of the present paper are awared of, the issue has remained as an open cosmological conundrum, with a few workers in the field having focused their attention on to those intriguing observations [35].

7. Appendix

A. Light propagation in NLED and birefringence

In this Appendix we shortly discuss the modification of the light velocity (birefringence effect) for the model of nonlinear electrodynamics $L(X)$. According to [39], one finds that the effect of birefringent light propagation in a generic model for nonlinear electrodynamics is given by the optical metric

$$g_1^{\mu\nu} = \mathcal{X}g^{\mu\nu} + (\mathcal{Y} + \sqrt{\mathcal{Y} - \mathcal{X}\mathcal{Z}})t^{\mu\nu}, \tag{76}$$

$$g_2^{\mu\nu} = \mathcal{X}g^{\mu\nu} + (\mathcal{Y} - \sqrt{\mathcal{Y} - \mathcal{X}\mathcal{Z}})t^{\mu\nu}, \tag{77}$$

where the quantities $\mathcal{X}$, $\mathcal{Y}$, and $\mathcal{Z}$ are related to the derivatives of the Lagrangian $L(X)$ and $t^{\mu\nu} = F^{\mu\alpha}F^{\nu}_{\alpha}$. For our model, expressed by Eq. (44), we infer ($K_1 = L_X, K_2 = 8L_{XX}$) [29]

$$g_1^{\mu\nu} = K_1(K_1 g^{\mu\nu} + 2K_2 t^{\mu\nu}), \quad g_2^{\mu\nu} = K_1^2 g^{\mu\nu}.$$

which show that birefringence is present. This means that some photons propagate along the standard null rays of spacetime metric $g^{\mu\nu}$, whereas other photons propagate along rays null with respect to the optical metric $K_1 g^{\mu\nu} + 2K_2 t^{\mu\nu}$.

The velocities of the light wave can be derived by using the light cone equations (effective metric). The value of the mean velocity, obtained by averaging over the directions of propagation and polarization, is given by

$$\langle v^2 \rangle \simeq 1 + (\delta - 1)R + (\delta - 1)^2 S, \tag{78}$$

$$R \equiv \frac{4}{3}\frac{T^{00}}{[4X + 2(\delta - 1)t^{00}]},$$

$$S = \frac{4}{3}\frac{\mathbf{S}^2}{[4X + 2(\delta - 1)t^{00}]^2},$$

where **S** is the energy flux density.

The high accuracy of optical experiments in laboratories requires tiny deviations from standard electrodynamics. This condition is satisfied provided $|\delta - 1| \ll 1$.

8. References

[1] M. Born, Nature (London) 132, 282 (1933); Proc. R. Soc. A 143, 410 (1934). M. Born, L. Infeld, Nature (London) 132, 970 (1933); Proc. R. Soc. A 144, 425 (1934). W. Heisenberg, H. Euler, Z. Phys. 98, 714 (1936). J. Schwinger, Phys. Rev. 82, 664 (1951).

[2] Bialynicka-Birula, Z. Bialynicki-Birula, I. (1970). Phys. Rev. D 2, 2341

[3] Blasi, P. & Olinto, A. V. (1999). Phys. Rev. D 59, 023001

[4] Burke, D. L., et al. (1997). Phys. Rev. Lett. 79, 1626

[5] Born, M. & Infeld, L. (1934). Proc. Roy. Soc. Lond. A 144, 425

[6] De Lorenci, V. A.., Klippert, R., Novello, M. & Salim, M., (2002). Phys. Lett. B 482, 134

[7] Delphenich, D. H. (2003). Nonlinear electrodynamics and QED. arXiv: hep-th/0309108

[8] Delphenich, D. H. (2006). Nonlinear optical analogies in quantum electrodynamics. arXiv: hep-th/0610088

[9] Denisov, V. I., Denisova, I. P. & Svertilov, S. I. (2001a). Doklady Physics, Vol. 46, 705.

[10] Denisov, V. I., Denisova, I. P., Svertilov, S. I. (2001b). Dokl. Akad. Nauk Serv. Fiz. 380, 435

[11] Denisov, V. I., Svertilov, S. I. (2003). Astron. Astrophys. 399, L39

[12] Dittrich, W. & Gies, H. (1998). Phys Rev.D 58, 025004

[13] Garcia, A.. & Plebanski, J. (1989). J. Math. Phys. 30, 2689

[14] Hadamard, J. (1903) Leçons sur la propagation des ondes et les equations de l'Hydrodynamique (Hermann, Paris, 1903)

[15] Following Hadamard [14], the surface of discontinuity of the EM field is denoted by Σ. The field is continuous when crossing Σ, while its first derivative presents a finite discontinuity. These properties are specified as follows: $\left[F_{\mu\nu}\right]_\Sigma = 0$, $\left[F_{\mu\nu|\lambda}\right]_\Sigma = f_{\mu\nu}k_\lambda$, where the symbol $\left[F_{\mu\nu}\right]_\Sigma = \lim_{\delta \to 0^+} (J|_{\Sigma+\delta} - J|_{\Sigma-\delta})$ represents the discontinuity of the arbitrary function J through the surface Σ. The tensor $f_{\mu\nu}$ is called the discontinuity of the field, $k_\lambda = \partial_\lambda\Sigma$ is the propagation vector, and symbols "$_|$" and "$_{||}$" stand for partial and covariant derivatives.

[16] Heisenberg, W. & Euler, H., (1936). Zeit. Phys. 98, 714

[17] Kandus, A.., Kerstin, K. E. & Tsagas, C. (2010). Primordial magneto-genesis. arXiv: 1007.3891 [astro-ph.CO]

[18] Landau, L. D. & Lifchiftz, E. (1970). Théorie des Champs. (Editions MIR, Moscou)

[19] Lichnerowicz, A., (1962). Elements of Tensor Calculus, (John Wiley and Sons, New York). See also Relativistic Hydrodynamics and Magnetohydrodynamics (W. A. Benjamin, 1967), and Magnetohydrodynamics: Waves and Shock Waves in Curved Space-Time (Kluwer, Springer, 1994)

[20] J.F. Plebanski, *Lectures on nonlinear electrodynamics*, monograph of the Niels Bohr Institute Nordita, Copenhagen (1968).

[21] M. Novello, S.E. Pérez Bergliaffa, J. Salim, Phys. Rev. D 69, 127301 (2004). See also V. A. De Lorenci et al., Phys. Lett. B 482, 134-140 (2000).

[??] H.J. Mosquera Cuesta, J.M. Salim, M. Novello, arXiv:0710.5188 [astro-ph]. L. Campanelli, P. Cea, G.L. Fogli, L. Tedesco, Phys. Rev. D 77, 043001 (2008); Phys. Rev. D 77, 123002

(2008). H. J. Mosquera Cuesta and J. M. Salim, MNRAS 354, L55 (2004). H. J. Mosquera Cuesta and J. M. Salim, ApJ 608, 925 (2004). H. J. Mosquera Cuesta, J. A. de Freitas Pacheco and J. M. Salim, IJMPA 21, 43 (2006) J-P. Mbelek, H. J. Mosquera Cuesta, M. Novello and J. M. Salim Eur. Phys. Letts. 77, 19001 (2007). J. P. Mbelek, H. J. Mosquera Cuesta, MNRAS 389, 199 (2008).

[23] H.J. Mosquera Cuesta, G. Lambiase, Phys. Rev. D 80, 023013 (20009).

[24] K.E. Kunze, Phys. Rev. D 77, 023530 (2008).

[25] M. Marklund, P.K. Shukla, Rev. Mod. Phys. 78, 591 (2006). J. Lundin, G. Brodin, M. Marklund, Phys. of Plasmas 13: 102102 (2006). E. Lundstrom *et al*. Phys. Rev. Lett. 96 083602 (2006).

[26] J-Q Xia, H. Li, X. Zhang, arXiv:0908.1876v2 [astro-ph.CO].

[27] G.V. Strotskii, Dokl. Akad. Nauk. SSSR 114, 73 (1957) [Sov. Phys. Dokl. 2, 226 (1957)].

[28] H. Pagels, E. Tomboulis, Nucl. Phys. B 143, 485 (1978).

[29] H. Mosquesta Cuesta and G. Lambiase, JCAP 1103 (2011) 033.

[30] J.D. Barrow, R. Marteens, Ch.G. Tsagas, Phys. Rep. 449, 131 (2007).

[31] M. Kamionkowski, A. Kosowski, A. Stebbins, Phys. Rev. D 55, 7368 (1997).

[32] G. Gubitosi, L. Pagano, G. Amelino-Camelia, A. Melchiorri, A. Cooray, JCAP 0908, 021 (2009). M. Das, S. Mohanty, A.R. Prasanna, arXiv:0908.0629v1[astro-ph.CO]. P. Cabella, P. Natoli, J. Silk, Phys. Rev. D 76, 123014 (2007). F.R. Urban, A.R. Zhitnitsky, arXiv:1011.2425 [astro-ph.CO]. G.-C. Liu, S. Lee, K.-W. Ng, Phys. Rev. Lett. 97, 161303 (2006). Y.-Z. Chu, D.M. Jacobs, Y. Ng, D. Starkman, Phys. Rev. D 82, 064022 (2010). S. di Serego Alighieri, arXiv:1011.4865 [astro-ph.CO]. R. Sung, P. Coles, arXiv:1004.0957v1[astro-ph.CO].

[33] M. Giovannini, Phys. Rev. D 81, 023003 (2010); Phys. Rev. D 80, 123013 (2009). M. Giovannini and K.E. Kunze, Phys. Rev. D 78, 023010 (2008). S. Alexander, J. Ochoa, and A. Kosowsky, Phys. Rev. D 79, 063524 (2009). M. Zarei, E. Bavarsad, M. Haghighat, R. Mohammadi, I. Motie, and Z. Rezaei, Phys. Rev. D 81, 084035 (2010). F. Finelli and M. Galaverni, Phys. Rev. D 79, 063002 (2009). A. Cooray, A. Melchiorri, and J. Silk, Phys. Lett. B 554, 1 (2003).

[34] I. Motie and S.-S. Xue, arXiv:1104.3555 [hep-ph].

[35] H. Mosquesra Cuesta and G. Lambiase, in preparation (2011).

[36] D. Hutsemekers, R. Cabanac, H. Lamy, D. Sluse, Astron.Astrophys. 441, 915 (2005).

[37] D. Hutsemekers, Astron. Astrophys 332, 410 (1998).

[38] D. Hutsemekers, H. Lamy, Astron. Astrophys 358, 835 (2000).

[39] Y.N. Obukhov, G.F. Rubilar, Phys. Rev. D 66, 024042 (2002).

8

Systematics in WMAP and Other CMB Missions

Hao Liu[1] and Ti-Pei Li[2]

[1]*Key Laboratory of Particle Astrophysics, Institute of High Energy Physics,*
Chinese Academy of Sciences
[2]*Department of Physics and Center for Astrophysics, Tsinghua University, Key Laboratory*
of Particle Astrophysics, Institute of High Energy Physics, Chinese Academy of Sciences
China

1. Introduction

1.1 The success of WMAP

When the WMAP team released their unprecedentedly precise CMB anisotropy maps in the year of 2003 (Fig. 1, Bennett et al. (2003)), everyone were celebrating the finally arrival of experimental foundation of precision cosmology. It did worth all praises, because the measured CMB power spectrum was so beautifully consistent with the theoretical expectation, and a clear scene of the birth and growth of a flat and ordered Universe seems to be right within the reach of our hands, as well as the first precision percentage estimation of its contents. Many people begin to believe that, we no longer need to be bored by one new cosmic model per day, and the Big Bang scene and ΛCDM model will be good enough to give us satisfactory explanation to anything we want to know.

It's not strange that there still remains some small disharmonious flaws, like an unexpectedly low quadrupole power (Bennett et al., 2003; Hinshaw et al., 2003b), alignment issues and NS asymmetry (Bielewicz et al., 2004; Copi et al., 2007; Eriksen et al., 2005; 2007; Hansen et al., 2004; 2006; Wiaux et al., 2006), non-Gaussianity including cold/hot spots (Copi et al., 2004; Cruz et al., 2005; 2007; Komatsu et al., 2003; Liu & Zhang, 2005; McEwen et al., 2006; Vielva et al., 2004; 2007), etc. Among all these flaws, the loss of the quadrupole power is the

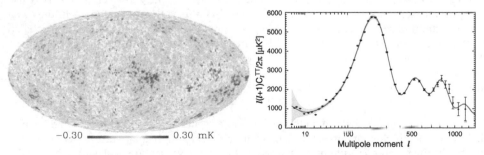

-0.30 ▬▬ ▬▬ 0.30 mK

Fig. 1. The CMB anisotropy map and power spectrum obtained by WMAP (Bennett et al., 2003).

sharpest one: We can see clearly from the right panel of Fig. 1 that the first black dot starting from the left (which represents the measured CMB quadrupole power) is much lower than theoretical expectation (the red line). However, in several years, these flaws or doubts don't really shake the success of WMAP, partly because they have a general difficulty: it's hard to distinguish the exact cause of these phenomenons, whether it's just something occasional, or due to a cosmic issue, or due to a measurement error. Given different answer to the cause, the corresponding treatment would be significantly different: For something occasional, nothing need to be done, and a cosmic issue need only to be solved in the frame of cosmic theories, but for a measurement error, it must be corrected by improving the measurement or data processing before considering any further theoretical efforts. Although, in several years, there are no strong evidences for us to find out the true cause, the cosmic issue is apparently preferred, because it's really difficult for a researcher outside the WMAP team to deal with the mission details. Especially, almost nobody really believes that the WMAP team, as such a large, experienced and professional group, could make a significant mistake in such an important experiment. Therefore, unless there appears something more evidently suggesting a potential measurement error, the effort of rechecking the WMAP detecting and data processing system will probably never be seriously considered, and the WMAP cosmology will never be questioned from the technical side.

1.2 Two early discovered anomalies

We had found such evidences between 2008 and 2009 (Li et al., 2009; Liu & Li, 2009a), and these findings had directly driven us to explore the WMAP raw data to find the reason for them. It's interesting that, till now we haven't found explanations that are satisfactory enough for the two early discovered anomalies, but instead, we have found anomalies that seem to be much more important than them.

1.2.1 The pixel-ring coupling issue

The first anomaly is related to a pixel-ring coupling issue on the CMB maps. To understand this, we need to know a few things about the way that the WMAP spacecraft measures the anisotropy of the CMB. The CMB anisotropy is extremely weak: about 10^{-5} of the well known nearly symmetric 2.73 K blackbody CMB emission. To detect such a weak signal, they have found a nice way to enhance the anisotropy relatively: The spacecraft receives CMB signals coming from two different directions by two highly symmetric antennas, and records only the difference between them to cancel the uniform 2.73 K blackbody CMB emission. This also help to counteract some of the systematical effects that affects the two antennas in the same way. In such an observational design, the separation angle between the two antennas is an important factor, which is $141°$ in the WMAP mission. It's apparently important to ensure that, when the measured differential signal is transformed into full-sky anisotropy maps by a sophisticated map-making process, the $141°$ separation angle, as an artificial thing, leaves completely no trace on the final map. In other words, there should be no clue for us to "guess" the man-made $141°$ separation angle from the final natural CMB map.

Intuitively, we would feel that, when the two antennas point at pixel A with a small deviation to 2.73 K (either higher or lower) and pixel B with very high temperature above 2.73 K respectively, the recorded differential data will have a large value; however, since we actually don't know either A or B temperature a priori, we will have at least two equally reasonable guesses: A is very cold or B is very hot. If the map-making from differential data is reliable,

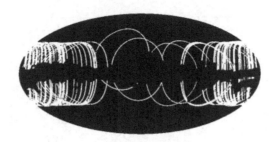

Fig. 2. The 141° scan rings of 2000 hottest pixels on the CMB sky.

then we would be able to deduce which is true with more observations that involves more pixels; however, if something doesn't work well, then we may obtain a wrong guess for A is very cold, or maybe not that bad: A is just slightly colder than what is expected. If such an deviation does exist, then it can be detected by checking the average temperature of all pixels in a ring that is 141° away from the very hot pixel B. This is right what we have done in Liu & Li (2009a).

In that work, we pick out 2000 hottest pixels from the CMB anisotropy map provided by the WMAP team, and select all pixels on the corresponding 141° rings which look like Fig. 2. The average temperature of these pixels are calculated for all WMAP bands Q, V and W, which are between -11 μK and -13 μK, and these values are 2.5 $\sim$ 2.7σ lower than expectation.

This is not enough yet: Although the values are really colder than expected, it can still be something occasional. However, the problem will become more serious if such a phenomena appear to be stuck on 141°, the man-made physical separation angle between the two antennas. It's possible to test this: Suppose that the final CMB map is perfect, then it must be completely "blind" to the physical separation angle, thus we can set a "guessed" value for the separation angle, then pick out each center pixel and the average temperature on the corresponding ring as a pair, and calculate the correlation coefficients between them to see if the 141° separation angle has an outstanding correlation strength. This is found to be true: By force the angular radius of the scan ring to change between 90° to 160°, we discovered that the anti-correlation strength is really strongest around 141°. Moreover, if the choice of center pixel is limited outside the foreground mask (so that the center pixel temperatures will not be very hot), then the correlation will be significantly weaker, indicating that these center pixels are less likely to arouse a cold ring effect (Fig. 3). With these self-consistent evidences, it's apparently more reasonable to deduce that the pixel-ring coupling is some kind of a systematical error, not something cosmic or occasional.

1.2.2 The T-N correlation issue

Another anomaly is much easier to understand: In any physical experiment, the most often adopted way to increase the accuracy and to suppress the noise is to apply more observations. With increasing number of observation, the result should be closer and closer to the true value, and converge at an accuracy level of $1/\sqrt{N_{obs}}$, but it's never expected that the result should subsequently increase or decrease with N_{obs}. In other words, there should be no correlation between the number of observation and the derived values. If this is seen with

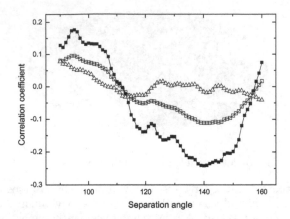

Fig. 3. The dependence of the correlation coefficients upon the separation angle. *Solid square*: The center pixels are limited within the hottest pixels on the CMB sky. *Empty square*: The center pixels being all pixel on the CMB sky. *Triangle*: The center pixels are limited outside the foreground mask.

enough significance, then it's almost certainly that a problem will exist in the measurement system, because the number of observation has nothing to do with the theoretical issues.

We did see this in the WMAP data (Li et al., 2009): the average absolute magnitude of the correlation coefficients between the WMAP CMB anisotropy temperatures and corresponding N_{obs} are $4.2\sigma \sim 4.8\sigma$ higher than expectation, and their distribution also significantly deviates from expected Gaussian distribution (Fig. 4), indicating that there is very likely a systematical problem in the WMAP mission, which has remarkably contaminated the final CMB anisotropy result.

When we see the river is muddy and believe it should not be like this, we will certainly go upstream to see where comes the mud. These two anomalies give rise to a doubt on the WMAP data, and will certainly drive us upstream for the origin, although the way is really cliffy.

2. Reprocessing the raw data

It's a tedious story how we explored the raw data and saw many false "differences" to the WMAP team. In brief, for many times when we worked on the WMAP raw data, we had seen this or that kind of "little surprises" smashed quickly by following tests, that we nearly despaired of finding anything worth noticing. This illustrated from another side how excellently had the WMAP team done. But at last we saw something that was not negligible on the problem of the CMB quadrupole. It's funny that it also started with a mistake: we actually got a much higher quadrupole value than WMAP at first, which was certainly welcome by theorist, but like usual, in a few days we found this is just another mistake. However, the turning point came silently before we realized it: When we corrected the mistake, the CMB quadrupole didn't come back and disappeared almost completely. We thought this was just one more fragile "little surprise", but we were wrong again.

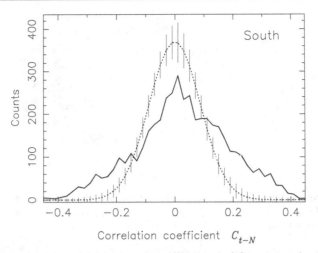

Fig. 4. The distribution of the T-N correlation coefficients (*solid*) compared with expectation and corresponding error bar given by 50,000 simulation (*dash*). It's very easy to see the deviation of real data from expectation.

2.1 Some basic thing about the map-making

In producing output full sky CMB temperature maps from the raw time-order differential data (TOD), the WMAP team have provided a beautiful formula (Hinshaw et al., 2003a): Let the CMB temperature anisotropy be a single-column matrix **T** with N_{pix} rows (for N_{pix} pixels on the sky), and the observation course be represented by a matrix **A** with N rows, N_{pix} columns (N is the total number of observations, which is much higher than N_{pix}) that is mostly zero, except for one 1 and one -1 in each row, then the TOD **D** can be described by a matrix multiplication:

$$\mathbf{AT} = \mathbf{D} \tag{1}$$

The jth line of Equation 1 is actually a simple sub-equation:

$$T_j^A - T_j^B = D_j, \tag{2}$$

where A and B stand for the two antennas, and $j = 1, \cdots, N$. The difficulty in solving for the CMB temperatures **T** is that **A** is not a square matrix. In linear algebra, this means either Equation 1 has no solution, or it contains many redundant rows, because N is much higher than N_{pix}. However, we can multiply the transpose of **A** to both side and obtain a square matrix $\mathbf{A}^T\mathbf{A}$:

$$(\mathbf{A}^T\mathbf{A})\mathbf{T} = \mathbf{A}^T\mathbf{D} \tag{3}$$

Then the solution can be formally obtained by:

$$\mathbf{T} = (\mathbf{A}^T\mathbf{A})^{-1}\mathbf{A}^T\mathbf{D} \tag{4}$$

Equation 4 seems to be simple and beautiful, but it's incorrect in linear algebra, because $\mathbf{A}^T\mathbf{A}$ is a singular matrix and $(\mathbf{A}^T\mathbf{A})^{-1}$ doesn't exist. Actually, even if $(\mathbf{A}^T\mathbf{A})^{-1}$ exists, there is no way to exactly solve for such a huge matrix with millions of rows and columns. An approximation

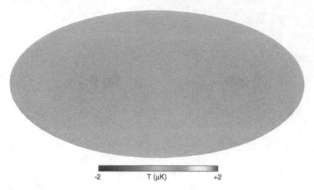

-2   T (µK)   +2

Fig. 5. The expected residual map-making error from Fig. 2 of Hinshaw et al. (2003a), which is negligible, and can be even reduced by more iterations. However, the observation noise is not taken into account here.

for $(\mathbf{A}^{\mathbf{T}}\mathbf{A})^{-1}$ is a $N_{pix} \times N_{pix}$ diagonal matrix[1]:

$$\mathbf{N}^{-1} = \begin{pmatrix} 1/N_1 & 0 & \cdots \\ 0 & 1/N_2 & \cdots \\ \cdots & \cdots & \cdots \end{pmatrix} , \tag{5}$$

where N_i is the total number of observation for the ith map pixel, $i = 1, \cdots, N_{pix}$ and $\sum_i N_i = N$. Based on this approximation, an iterative solution can be preformed to estimate the final CMB temperatures $\mathbf{T}$.

The iterative solution is simple: we give an initial all-zero guess $\mathbf{T}^0$ to the CMB anisotropy map $\mathbf{T}$, and, based on that, use the TOD $\mathbf{D}$ to improve the guess by $T_j^{1,i_A} = T_j^{0,i_B} + D_j$ or $T_j^{1,i_B} = T_j^{0,i_A} - D_j$ (see also Equation 2), where T_j means the jth temperature estimation corresponding to the jth observation, and i_A stands for the A-side pixel in this observation, so do i_B. In this way, each D_j gives two improved estimations T_j^{1,i_A} and T_j^{1,i_B} for the positive/negative sides one the map respectively. The improved $\mathbf{T}^1$ is the assembly of all T_j, averaged at each sky pixel i respectively: $T_i^1 = (\sum_j T_j^{1,i_A=i} + \sum_j T_j^{1,i_B=i})/N_i$ (T_i mean the averaged temperature of the ith pixel on the sky), and the final CMB anisotropy map is produced by 50 to 80 iterations like this[2].

Although a strict solution to $\mathbf{T}$ is impossible, the iterative solution works quite well, at least according to simulation: As presented by the WMAP team (Fig. 5), discarding the observation noise and the uncertain monopole, the residual full sky map-making error is negligible. This has been confirmed by us, but we have found that for such an excellent convergence in Fig. 5, the average of odd and even iteration rounds should be used, which was not mentioned in the WMAP documents.

---

[1] Since $(\mathbf{A}^{\mathbf{T}}\mathbf{A})^{-1}$ doesn't exist, "approximation for $(\mathbf{A}^{\mathbf{T}}\mathbf{A})^{-1}$" means this matrix multiplies $\mathbf{A}^{\mathbf{T}}\mathbf{A}$ gives an output matrix that is almost unitary.

[2] The map-making processes described here are simplified for ideally symmetric antennas with perfectly stable responses. For equations including necessary fine corrections, please refer to Hinshaw et al. (2003a)

2.2 Be consistent to WMAP first

Although our emphasis is the difference to the WMAP team, we would like to illustrate at the very beginning that we are now able to obtain fully consistent results to the WMAP team using our own map-making software, as shown in Fig 6. The reason is: Unless we are able to do so, the crucial reason of the **difference** between our results and WMAP will never be determined, because we will be lost in countless technical details, each seems to be able to cause some specious "differences" in this or that way, as briefly but incompletely listed below:

- f^{-1} noise (an equipment feature)
- Usage of the processing mask (determines how many raw data are unused)
- TOD flag issue (likewise)
- Antenna imbalance (the antennas are not absolutely symmetric)
- Problem in resolving the spacecraft velocity (affects the Doppler signal and calibration)
- The Sun velocity uncertainty (likewise)
- Dipole signal subtraction (remove the unwanted Doppler signal)
- Foreground subtraction (remove the unwanted foreground emission)
- Incomplete sky coverage (some sky regions are unused in calculating the CMB power spectrum)
- Beam function correction (which can greatly suppress the small scale anisotropy)
- Window function correction (likewise)
- Map making convergence (does the iteration converges well?)
- The antenna pointing vectors (affected by various factors)
- . . .

Needless to say, nothing valuable can be obtained before we climb out of such a bottomless list. However, once we get Fig. 6, we will then be able to clearly identify which item in this list makes the major contribution to the difference between WMAP and us. This can be done by changing one item one time, and see how it takes effect on the final CMB maps and power spectrum.

3. The origin of the difference between our results and WMAP

In time sequence, the difference in the CMB anisotropy maps and power spectrum is discovered first, and then the origin of the difference is found and confirmed as illustrated in Section 2.2. This is not the end of the problem, because we need to check whether WMAP or us is more likely to be correct. Soon after that, several evidences supporting our results are obtained, and we also found reasonable explanations to why WMAP could be wrong. Such explanations is not only valuable in improving the WMAP result, but also valuable in preventing future CMB detecting mission from making the same mistake. We will follow this sequence, and present our work step by step.

3.1 The difference in the CMB result using the same raw data

After tottering through the raw data processing, we finally get a tentative CMB result in the end of 2008 (Liu & Li, 2009b), which looks very similar to the WMAP official release, as shown in Fig. 7 However, if we subtract our map from WMAP and smooth the result, the difference can be clearly seen, which is a four-spot structure, two hot and two cold. This is a typical quadrupole structure. More interestingly, such a difference is almost the same to the claimed

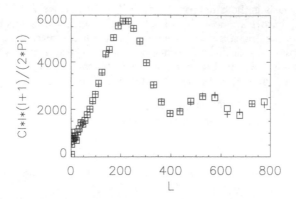

Fig. 6. We are now able to obtain fully consistent CMB result to the WMAP team using our own software but adopting their convention. This greatly simplify the search for the origin of the difference. *Square*: CMB power spectrum obtained from the WMAP official maps. *Cross*: CMB power spectrum obtained from our maps using their conventions.

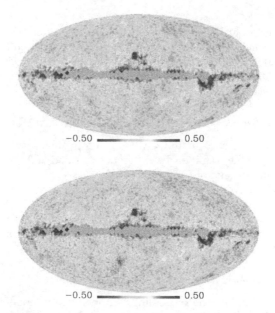

Fig. 7. The Q1-band CMB temperature maps by the WMAP team (*left panel*) and by us (*right panel*). Both in the Galactic coordinate and the unit is mK. They look almost the same.

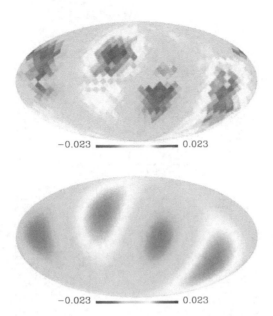

Fig. 8. This figure is to show the large-scale CMB structure difference between WMAP and us. *Left panel:* The CMB map of WMAP minus ours, smoothed to $N_{side} = 8$, so that it's dominated by large-scale differences. *Right panel:* The quadrupole component of the released WMAP CMB anisotropy map. They are amazingly similar, which casts doubt on the WMAP data.

CMB quadrupole structure by WMAP (Fig. 8). This phenomena certainly worths further consideration.

The CMB power spectrum is calculated by spherical harmonic decomposition. The spherical harmonics $Y_{lm}(\theta, \phi)$ are known as a family of normalized, orthogonal and complete functions on the sphere, which is widely used to analyze problems on the sphere including the CMB anisotropy. The spherical harmonic coefficients are calculated by:

$$\alpha_{lm} = \int T(\theta, \phi) Y_{lm}^*(\theta, \phi) \sin(\theta) d\theta d\phi, \tag{6}$$

and the CMB power spectrum is obtained by:

$$C_l = \frac{1}{2l + 1} \sum_{m=-l}^{l} |\alpha_{lm}|^2 \tag{7}$$

Besides Equation 7, there are many further corrections in calculating the final CMB power spectrum; however, as illustrated in Section 2.2, we can confirm that we have appropriately applied all these corrections by the consistency between the crosses/squares in Fig. 6 (so we see how important it is to be consistent to WMAP first). Based on the confirmed consistency, we can conclude that the differences in both large and small scale power spectra (Fig. 9) are

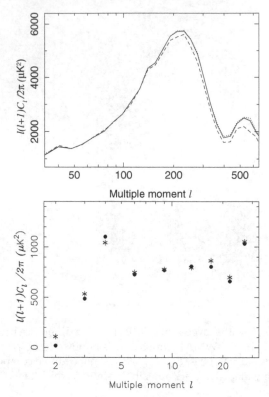

Fig. 9. *Left Panel:*The CMB power spectra derived with our software from our new map (*dash line*), with our software but from the WMAP official maps (*dotted line*), and directly released by the WMAP team (*solid line*). By the difference between the solid line and dash line, we can see that the small scale CMB power spectra derived by us and WMAP are apparently different. *Right Panel:* The large scale CMB power spectra by us (*dot*) and by WMAP (*asterisk*). We can see that only the quadrupole is significantly different.

not due to the power spectra estimation, but due to a possible issue in the map-making from the raw data.

3.2 The origin of the difference

We took a long time to search for the origin of the differences. Finally we found that it's due to an antenna pointing vector issue. This had been noticed by us before, but was ignored at first because it's really too small.

The WMAP spacecraft doesn't record the antenna pointing vector for each observation: The pointing vectors are much less frequently recorded than the science data in order to reduce the data file size. Therefore in raw data processing, we need to interpolate the available antenna pointing vector to calculate the unrecorded pointing vectors for each observation. In our data processing program, one of the interpolation parameters is slightly different to WMAP, which causes the calculated pointing vectors to be different for about half a pixel to them. This looks really a trivial thing, but it's right the key of all differences shown in Fig. 8 and 9.

4. Zigzag

In most cases, when you find the origin of the problem, the case will soon be totally solved. However, for us, discovering the antenna pointing vector issue is just the beginning of the zigzag pursuit.

4.1 WMAP seem to be right?

When we study the key of the difference: the interpolation parameter change carefully, we feel that WMAP is probably correct, because they have used the center of each observation interval as the effective antenna pointing vector for that observation, but we have used the start of it. Since WMAP is continuously receiving the microwave signal, the center of the observation is apparently a better choice. Thus the prospect of our finding seem to be dim at first.

4.2 A doubtful point: evidence against WMAP

However, an unconvinced fact still remain in our mind: When adopting our interpolating settings, nearly 90% of the CMB quadrupole (the component with $l = 2$) will disappear. If this doesn't look suspicious enough, then from another point of view we will see what it actually means.

The antenna pointing vector difference affects the CMB quadrupole via the Doppler signal subtraction. In the WMAP observation, the strongest contamination to the CMB signal is the Doppler signal caused by the motion of the spacecraft towards the CMB test frame. This particular signal can be calculated by:

$$d = \frac{T_0}{c} \mathbf{V} \cdot (\mathbf{n_A} - \mathbf{n_B}),\tag{8}$$

where T_0 is the 2.73 K CMB monopole, c is the speed of light in vacuum, $\mathbf{V}$ is the velocity of the spacecraft relative to the CMB rest frame, and $\mathbf{n_A}$, $\mathbf{n_B}$ are the antenna pointing vectors. We can see clearly that, if the antenna pointing vectors are slightly different, either on $\mathbf{n_A}$ or $\mathbf{n_B}$ or both, then the calculated Doppler dipole signal d will be consequently different:

$$\Delta d = \frac{T_0}{c} \mathbf{V} \cdot \Lambda \mathbf{n}, \quad \Delta \mathbf{n} = \Delta \mathbf{n_A} - \Delta \mathbf{n_B}.\tag{9}$$

It's interesting that, from Equation 9 we can see that we don't have to know any CMB information in calculating Δd. Therefore, the deviation upon the CMB quadrupole caused by possible antenna pointing vector error (no matter what reason) can be calculated independently of the CMB maps or CMB detection. This has been done by us (Liu, Xiong & Li 2010), and the result is again almost same to the released WMAP CMB quadrupole, as shown in Fig. 10. Now we can see the real puzzle: The "CMB" quadrupole has been reproduced without any observation, which is absurd. Theoretically speaking, this is not impossible, but a much more reasonable explanation is that the claimed "CMB" quadrupole is actually a systematical error.

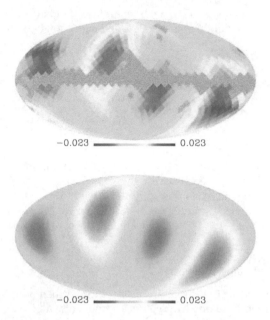

Fig. 10. *Left Panel:* The expected deviation on final CMB temperature map caused by Δd (see Equation 9). Note that there is no inclusion of any CMB signal in obtaining this figure. *Right Panel:* The claimed CMB quadrupole by WMAP.

The results in Fig. 10 has been independently reproduced by Moss et al. (2010) and Roukema (2010a), and has been regarded as an important evidence in questioning the WMAP cosmology (Sawangwit & Shanks, 2010). The software used to obtain this result is the same software written by us for re-processing the WMAP TOD, which is publicly available on the websites of the cosmocoffee forum [3] or the Tsinghua Center for Astrophysics[4].

4.3 WMAP is supported again?

Since the focus of the problems seem to be upon the antenna pointing vectors, a third party test on this is soon applied by Roukema (2010a), in which he believe that, if there is an antenna pointing vector error, then the generated CMB maps will be blurred, thus by checking the image sharpness of the CMB maps, we can decide whether there really exists such an error. His conclusion is that, the WMAP official maps are apparently less affected by the blurring effect compared to ours. This is a strong supporting evidence for WMAP being correct, and the prospect of our finding seems to be dim again.

4.4 Not the end: more details discovered

In fact, even in Roukema (2010a), the author didn't conclude that there will surely be no problem in the pointing vectors. He pointed out that the antenna pointing vector issue might

[3] http://cosmocoffee.info/viewtopic.php?t=1541
[4] http://dpc.aire.org.cn/data/wmap/09072731/release_v1/source_code/v1/

still take effect in the data calibration. Although such an possibility seems to be minor, it was soon confirmed by following works by him and us.

Three months later after Roukema's previous work, by checking the median per map of the temperature fluctuation variance per pixel, he discovered that, in a very high significance level, there does seem to be an antenna pointing vector error in the raw data, possibly due to the data calibration (Roukema, 2010b). We have confirmed this finding with different method (Liu, Xiong & Li, 2011), and the obtained results are very well consistent to Roukema (2010b). This seems to be puzzling: How can the same problem, the antenna pointing issue, be confirmed and rejected at the same time?

4.5 Real and equivalent pointing vector errors

To answer this question, we need to have a look at the antenna pointing issues again. The antenna directions are affected by at least three independent factors:

- *Direct pointing error (e.g., the antennas are misplaced).*
- *Timing error*
- *Sidelobe uncertainty*

All these three effects can cause antenna pointing vector errors; however, we have confirmed that the 1st and 2nd effects can further cause both the blurring effect and power spectrum deviation, and the 3nd causes only the power spectrum deviation. Thus the evidences found in Roukema (2010a) doesn't in principle conflict with Roukema (2010b) and Liu, Xiong & Li (2011). Moreover, all these three articles have also pointed out that the antenna pointing error can possibly take effect in the calibration stage without blurring the final image, which lead to the same conclusion.

Now we have a clear logical sequence: The three reasons listed above (possibly more) cause antenna pointing vector errors (real or equivalent), and the antenna pointing vector errors further cause all the CMB deviations discussed in this article. The antenna pointing vector error is hence the node of the entire problem.

4.5.1 Real pointing vector error

It's easy to understand the first factor, thus we introduce no more discussion for it. The second factor is due to the rotation of the spacecraft: In order to scan the sky, the spacecraft must rotate continuously, thus we must know the angular velocity and the local time to calculate the antenna pointing vectors. If there is a timing error, then the derived antenna pointing vectors will certainly be mistaken. In Liu, Xiong & Li (2010), we have discovered that the WMAP spacecraft attitude data are asynchronous to the CMB differential data, thus existence of the timing error is apparently possible. This is further confirmed in Liu, Xiong & Li (2011) and Roukema (2010b).

Both the first and the second factors are called real pointing vector errors, because the antenna pointing vectors will be substantially in error due to them. However, besides the real pointing vector error, there is also equivalent pointing vector error caused by the third factor. In this case, the antenna pointing vectors might be accurate, but the recorded data are twisted, as if there were a pointing vector error.

4.5.2 Equivalent pointing vector error

The third factor is a hidden factor that has never been noticed before. Generally speaking, radio antenna has response to all 4π solid angle, not merely along it's optical pointing. For the WMAP antennas, the 4π response is described by a normalized gain G: If there is only one beam b coming from a direction marked by pixel i, then the recorded signal should be $S = G_i b$. For full sky signal like the CMB, the recorded signal is[5]:

$$S = \frac{1}{N_{pix}} \sum_{i=0}^{N_{pix}-1} G_i T_i . \tag{10}$$

Since WMAP works in differential mode, the equation for WMAP should be:

$$D = \frac{1}{N_{pix}} \sum_{i=0}^{N_{pix}-1} (G_i^A - G_i^B) T_i , \tag{11}$$

where N_{pix} is the number of pixels on the sky, G_i^A and G_i^B are the normalized gains for the A-side and B-side antennas in the spacecraft coordinate[6] respectively, T_i is the CMB temperature. According to the WMAP convention, the normalization rule is $\sum_{i=0}^{N_{pix}-1} G_i = N_{pix}$.

In Equation 11, only a few pixels stands for the main lobe, and most pixels belong to the sidelobe, but the sidelobe pixels have much lower gain amplitudes: the summation of all sidelobe normalized gains are less than 5%. The standard method to clean the sidelobe contamination is by deconvolution; however, this is very complex and slow, and the biggest disadvantage is missing of a clear physical picture. We discovered that the deconvolution can be greatly simplified for some special full sky signals, and we can get a very clear and simple physical scene for them. Fortunately, the strongest contamination in CMB experiments, the Doppler signal, is right such a kind of signal.

According to Equation 8 and Equation 11, we can calculate the 4π response to the Doppler signal by

$$d_{sidelobe} = \frac{T_0}{c} \mathbf{v} \cdot \sum_k \left[\frac{(G_k^A - G_k^B)}{N} \mathbf{n_k} \right] , \tag{12}$$

where k stands for all sidelobe pixels.

Using ΔG^A and ΔG^B for the sidelobe gain uncertainties, and considering the fact that G^A and G^B are both constants in the spacecraft coordinate, we consequently obtain two constants in the spacecraft coordinate:

$$\Delta \mathbf{n_A} = \sum_k \frac{\Delta G_k^A \mathbf{n_k}}{N} , \quad \Delta \mathbf{n_B} = \sum_k \frac{\Delta G_k^B \mathbf{n_k}}{N} , \tag{13}$$

---

[5] The transmission factors are omitted here, e.g., the ratio between the antenna temperature and CMB temperature, and the ratio between temperature and electronic digital unit. That means, with all these simplifications, if the antenna response is a δ function which is none zero only at the optical direction, then the recorded signal is simply $S = b$

[6] Because the antennas are static in the spacecraft coordinate.

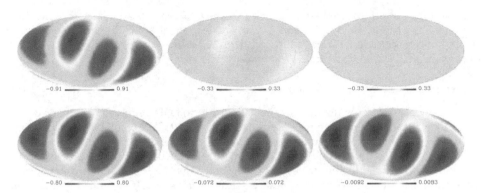

Fig. 11. *Top Panels:* The expected temperature deviation caused by the equivalent pointing vector error due to the sidelobe uncertainty issue in the WMAP mission. *Bottom panels:* The quadrupole components of the top panels. From left to right: The equivalent pointing vector error is along the x, y, z axes in the spacecraft coordinate respectively.

and the uncertainty of $d_{sidelobe}$ can be expressed as

$$\Delta d_{sidelobe} = \frac{T_0}{c}\mathbf{v} \cdot (\Delta\mathbf{n_A} - \Delta\mathbf{n_B}) = \frac{T_0}{c}\mathbf{v} \cdot \Delta\mathbf{n}. \tag{14}$$

$\Delta\mathbf{n}$ is called the equivalent pointing vector error, because it isn't a real pointing error, but has the same effect as real pointing vector errors caused by direct reasons or the timing issue.

It's apparent that we are interested in the gain uncertainty ΔG, not the gain itself. It's very difficult to accurately determine the antenna sidelobe response, at least due to three reasons: 1, the sidelobe response is very weak; 2, the signal due to the sidelobe response is always mixed with other much stronger signals (some are even unknown); 3, the sidelobe response of the two antennas could overlap and it's almost impossible to exactly solve for both. Therefore, it's not strange that the WMAP sidelobe gain has very high uncertainty: as presented by Barnes et al. (2003), up to 30%; however, the provides 30% uncertainty contains only the average level uncertainty, with no inclusion of the pixel-to-pixel variance. Thus the overall uncertainty must be much higher than the claimed 30%.

It's well known that the dot product of two vectors is invariant in coordinate transform, thus we can calculate $\Delta d_{sidelobe}$ in any coordinates using Equation 14, and the result will be the same. Since $\Delta\mathbf{n}$ is determined by the gain and antenna pointing vectors, which are both constants in the spacecraft coordinate, the best coordinate to calculate $\Delta d_{sidelobe}$ is certainly the spacecraft coordinate. However, even in the spacecraft coordinate, it's still impossible to exactly calculate the uncertainty due to $\Delta\mathbf{n}$ (otherwise it won't be called the "uncertainty"). In this case, a possible way out is to divide $\Delta\mathbf{n}$ into three components Δn_x, Δn_y, Δn_z along the X, Y, Z axes of the spacecraft coordinate. For each component, we can set any amplitude for it (we will explain why we can do like this below) and use the corresponding $\Delta d_{sidelobe}$ instead of the differential data to obtain an output map ΔT_x, ΔT_y or ΔT_z like we did for the real TOD in Sec. 2.1. Examples of ΔT_x, ΔT_y, ΔT_z are presented in Fig. 11.

Both Fig. 10 and Fig. 11 illustrate the same thing: there could be significant systematical error in the WMAP CMB detection. Fig. 11 tell us more, that even if the antenna pointing vectors are accurate, it's still impossible to conclude that the CMB results are hence reliable. Not to mention the fact confirmed by Roukema (2010b) and Liu, Xiong & Li (2011) that there are more potential uncertainties in the data calibration. Thus it's really too early to be so sure about the CMB detection or a final CMB model right now.

4.6 How to remove the artificial components?

We have seen possible CMB temperature deviations due to real or equivalent antenna pointing errors in Fig 10, 11, but they don't 100% ensure that the final CMB result by WMAP is wrong. However, we can still do something to get a more reasonable CMB estimation. We know that if there is extra uncertainty due to any reason in CMB detection, the corresponding deviation map, if available, should be removed from the final CMB temperature map to ensure a clean and reliable result. Now we have successfully obtained the final temperature deviation pattern due to the possible pointing error (Fig. 10 or Fig. 11, no matter real or equivalent), then we surely need to remove it from the obtained CMB results. The last hamper is that we don't know the exact amplitude of Δn_x, Δn_y, Δn_z; however, there are already methods for that, which has already been adopted by the WMAP team in removing the foreground emission.

When we try to remove the foreground emission, we are facing exactly the same difficulty: The foreground emission can be modeled by astrophysical emission mechanism including free-free, synchrotron, and dust emissions, however, the exact amplitude of each emission mechanism can't be predicted by the model. Moreover, all emission mechanism take effect together in CMB detection, and they are also combined with the CMB signal, thus the only way to determine the amplitudes of each emission mechanism is by model fitting. In this process, three a priori emission maps are presented as T_{ff}, T_{sync}, and T_{dust} (Bennett et al., 2003; Finkbeiner et al., 1999; Finkbeiner, 2003; Gold et al., 2009), and the clean temperature T is supposed to be

$$T = T^* - c_{ff}T_{ff} - c_{sync}T_{sync} - c_{dust}T_{dust}, \tag{15}$$

where the amplitudes c_{ff}, c_{sync}, c_{dust} are calculated by least-square fitting. The same technic can be used here: The real CMB temperature should be

$$T = T^* - c_x\Delta T_x - c_y\Delta T_y - c_z\Delta T_z, \tag{16}$$

and like above, the coefficients c_x, c_y, c_z can be determined by least-square fitting. It's important to notice that, in least square fitting, the amplitudes of the templates ΔT_x, ΔT_y, ΔT_z are not important, that's why we said above that we can use any amplitude for Δn_x, Δn_y, and Δn_z. As shown by Liu & Li (2010), the result is well consistent to our previous work (Liu & Li, 2009b): the CMB quadrupole after template fitting removal is about $10 \sim 20~\mu K$, much lower than the WMAP release. In other words, without supposing an a priori pointing error, we have self-consistently confirmed it posteriorly.

5. Planck and future CMB detection missions

Although the Planck spacecraft is significantly improved compared to WMAP, the basic detecting method are similar: antenna that suffers from 4π sidelobe response is used; the spacecraft works on the same L2 spot; the scan pattern is similar: a three-axis rotation consists of rotation around the Sun, spin around it's symmetry axis, and cycloidal procession around the Sun-to-spacecraft axis. Thus the three factors we have listed: direct pointing error, timing error, and sidelobe uncertainty can all take effect in the Planck mission. We have simulated the

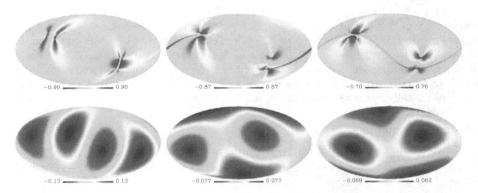

Fig. 12. *Top Panels:* The expected temperature deviation caused by the equivalent pointing vector error due to the sidelobe uncertainty issue in the Planck mission. *Bottom panels:* The quadrupole components of the top panels. From left to right: The equivalent pointing vector error is along the x, y, z axes in the spacecraft coordinate respectively.

Planck scan strategy and possible resulting CMB deviation due to real or equivalent pointing vector error (Liu & Li, 2011), and found that similar CMB deviation may also occur in the Planck mission, as shown in Fig. 12. If in the final data release of Planck, the CMB result after least-square fitting removal is consistent to the WMAP result with the similar treatment, then the corresponding result would certainly be more robust and import.

6. Discussion

Understanding systematic effects in experiments and observations is a difficult task. It is not strange for a newly explored waveband in astronomy that early observation results contain unaware systematics. As an example, COS B is the second mission for gamma-ray band. The COS B group published the 2CG catalog of high energy gamma-ray sources detected with the selection criteria for no more than one spurious detection above the adopted threshold from background fluctuation could be expected (Swanenburg et al., 1981). However, Li & Wolfendale (1982) found that about half of the released 25 sources are pseudo-ones produced by Galactic gamma-ray background fluctuation. The mistake came from a simplified model of the diffuse Galactic gamma-rays used by the COS B group, leading to systematically overestimate significance of source detection. After many years debating, the pseudo-sources have been finally deleted from the 2CG catalog, and instead of the simplified background model, more reliable structured models for diffuse Galactic gamma-rays have been adopted by all following gamma-ray missions. For microwave band, WMAP is just the second mission. Although the WMAP team made huge effort to study systematic effects, there still notably exist systematic errors in their released results. It seems that the foreground issue has been carefully and effectively treated, but the CMB dipole effect has not yet. The Doppler-dipole moment dominates CMB anisotropies, various errors in CMB experiments can produce pseudo-dipole signals and then artificial anisotropies in resultant CMB maps, seriously affecting cosmological studies. Beside the foreground emissions, the pseudo-dipole-induced anisotropy is another key systematic problem common for all CMB missions. Without carefully and effectively treating the dipole issue, CMB maps from COBE, WMAP, and Planck are not reliable for studying the CMB anisotropy.

However, we still have to notice one thing: we have confirmed that the early discovered anomalies (Sec. 1) still exist after least-square fitting removal, although they do seem to be weakened. Another fact also need to be noticed: Both the WMAP CMB quadrupole and octupole is aligned with the Ecliptic plane and each other (Bielewicz et al., 2004; Copi et al., 2004), however, after least-square fitting removal, the quadrupole no longer aligned with either one, but the octupole still aligned with the Ecliptic plane. This two facts may indicate that there are still unresolved problems in the WMAP mission!

7. Acknowledgments

This work is Supported by the National Natural Science Foundation of China (Grant No. 11033003). The data analysis made use of the WMAP data archive and the HEALPix software package.

8. References

Barnes C., et al. (2003). First-Year Wilkinson Microwave Anisotropy Probe (WMAP) Observations: Galactic Signal Contamination from Sidelobe Pickup, *The Astrophysical Journal Supplement Series*, 148, 51-62

Bennett C. L., et al. (2003). First-Year Wilkinson Microwave Anisotropy Probe (WMAP) Observations: Preliminary Maps and Basic Results, *The Astrophysical Journal Supplement Series*, 148, 1-27

Bielewicz P., Górski K. M., Banday A. J. (2004). Low-order multipole maps of cosmic microwave background anisotropy derived from WMAP, *Monthly Notices of the Royal Astronomical Society*, 355, 1283-1302

Copi C. J., Huterer D., Starkman G. D. (2004). Multipole vectors: A new representation of the CMB sky and evidence for statistical anisotropy or non-Gaussianity at $2 \leq l \leq 8$, *Physical Review D*, 70, 043515-

Copi C. J., Huterer D., Schwarz D. J., Starkman G. D. (2007). Uncorrelated universe: Statistical anisotropy and the vanishing angular correlation function in WMAP years 1 3, *Physical Review D*, 75, 023507-

Cruz M., Martínez-González E., Vielva P., Cayón L. (2005). Detection of a non-Gaussian spot in WMAP, *Monthly Notices of the Royal Astronomical Society*, 356, 29-40

Cruz M., Cayón L., Martínez-González E., Vielva P., Jin J. (2007). The Non-Gaussian Cold Spot in the 3 Year Wilkinson Microwave Anisotropy Probe Data, *The Astrophysical Journal*, 655, 11-20

Eriksen H. K., Banday A. J., Górski K. M., Lilje P. B. (2005). The N-Point Correlation Functions of the First-Year Wilkinson Microwave Anisotropy Probe Sky Maps, *The Astrophysical Journal*, 622, 58-71

Eriksen H. K., Banday A. J., Górski K. M., Hansen F. K., Lilje P. B. (2007). Hemispherical Power Asymmetry in the Third-Year Wilkinson Microwave Anisotropy Probe Sky Maps, *The Astrophysical Journal*, 660, L81-L84

Finkbeiner D. P., Davis M., Schlegel D. J. (1999). Extrapolation of Galactic Dust Emission at 100 Microns to Cosmic Microwave Background Radiation Frequencies Using FIRAS, *The Astrophysical Journal*, 524, 867-886

Finkbeiner D. P. (2003). A Full-Sky Hα; Template for Microwave Foreground Prediction, *The Astrophysical Journal Supplement Series*, 146, 407-415

Gold B., et al. (2009). Five-Year Wilkinson Microwave Anisotropy Probe Observations: Galactic Foreground Emission, *The Astrophysical Journal Supplement Series*, 180, 265-282

Hansen F. K., Cabella P., Marinucci D., Vittorio N. (2004). Asymmetries in the Local Curvature of the Wilkinson Microwave Anisotropy Probe Data, *The Astrophysical Journal*, 607, L67-L70

Hansen F. K., Banday A. J., Eriksen H. K., Górski K. M., Lilje P. B. (2006). Foreground Subtraction of Cosmic Microwave Background Maps Using WI-FIT (Wavelet-Based High-Resolution Fitting of Internal Templates), *The Astrophysical Journal*, 648, 784-796

Hinshaw G., et al. (2003). First-Year Wilkinson Microwave Anisotropy Probe (WMAP) Observations: Data Processing Methods and Systematic Error Limits, *The Astrophysical Journal Supplement Series*, 148, 63-95

Hinshaw G., et al. (2003). First-Year Wilkinson Microwave Anisotropy Probe (WMAP) Observations: The Angular Power Spectrum, *The Astrophysical Journal Supplement Series*, 148, 135-159

Komatsu E., et al. (2003). First-Year Wilkinson Microwave Anisotropy Probe (WMAP) Observations: Tests of Gaussianity, *The Astrophysical Journal Supplement Series*, 148, 119-134

Li T.P., Wolfendale A.W. (1982). Discret sources of cosmic gamma rays, *Astronomy and Astrophysics*, 116, 95-100

Li T.P., Liu H., Song L.-M., Xiong S.-L., Nie J.-Y. (2009). Observation number correlation in WMAP data, *Monthly Notices of the Royal Astronomical Society*, 398, 47-52

Liu H., Li T.P. (2009). Systematic distortion in cosmic microwave background maps, *Science in China G: Physics and Astronomy*, 52, 804-808

Liu H., Li T.P. (2009). Improved CMB Map from WMAP Data, *ArXiv e-prints*, arXiv:0907.2731

Liu H., Xiong S.L., Li T.P. (2010). The origin of the WMAP quadrupole, *ArXiv e-prints*, arXiv:1003.1073

Liu H., Li T. P. (2010). Pseudo-Dipole Signal Removal from WMAP Data, *Chinese Science Bulletin*, 56(1) 29-33

Liu H., Xiong S.L., Li T.P. (2011). Diagnosing timing error in WMAP data, *Monthly Notices of the Royal Astronomical Society*, 413, L96-L100

Liu H., Li T.P. (2011). Observational Scan-Induced Artificial Cosmic Microwave Background Anisotropy, *The Astrophysical Journal*, 732, 125

Liu X., Zhang S. N. (2005). Non-Gaussianity Due to Possible Residual Foreground Signals in Wilkinson Microwave Anistropy Probe First-Year Data Using Spherical Wavelet Approaches, *The Astrophysical Journal*, 633, 542-551

McEwen J. D., Hobson M. P., Lasenby A. N., Mortlock D. J. (2006). A high-significance detection of non-Gaussianity in the WMAP 3-yr data using directional spherical wavelets, *Monthly Notices of the Royal Astronomical Society*, 371, L50-L54

Moss A., Scott D., Zibin J. P. (2010). No evidence for anomalously low variance circles on the sky, *ArXiv e-prints*, arXiv:1012.1305

Roukema B. F. (2010). On the suspected timing error in Wilkinson microwave anisotropy probe map-making, *Astronomy and Astrophysics*, 518, A34

Roukema B. F. (2010). On the suspected timing-offset-induced calibration error in the Wilkinson microwave anisotropy probe time-ordered data, *ArXiv e-prints*, arXiv:1007.5307

Sawangwit, U. & Shanks, T. (2010). Is everything we know about the universe wrong? *Astronomy & Geophysics*, 51: 5.14Í C5.16

Swanenburg B.N. et al., (1981). Second COS B catalog of high energy gamma-ray sources, *The Astrophysical Journal*, 243, L69

Vielva P., Martínez-González E., Barreiro R. B., Sanz J. L., Cayn L. (2004). Detection of Non-Gaussianity in the Wilkinson Microwave Anisotropy Probe First-Year Data Using Spherical Wavelets, *The Astrophysical Journal*, 609, 22-34

Vielva P., Wiaux Y., Martínez-González E., Vandergheynst P. (2007). Alignment and signed-intensity anomalies in Wilkinson Microwave Anisotropy Probe data, *Monthly Notices of the Royal Astronomical Society*, 381, 932-942

Wiaux Y., Vielva P., Martínez-González E., Vandergheynst P. (2006). Global Universe Anisotropy Probed by the Alignment of Structures in the Cosmic Microwave Background, *Physical Review Letters*, 96, 151303-

Permissions

The contributors of this book come from diverse backgrounds, making this book a truly international effort. This book will bring forth new frontiers with its revolutionizing research information and detailed analysis of the nascent developments around the world.

We would like to thank Herman J. Mosquera Cuesta , for lending his expertise to make the book truly unique. He has played a crucial role in the development of this book. Without his invaluable contribution this book wouldn't have been possible. He has made vital efforts to compile up to date information on the varied aspects of this subject to make this book a valuable addition to the collection of many professionals and students.

This book was conceptualized with the vision of imparting up-to-date information and advanced data in this field. To ensure the same, a matchless editorial board was set up. Every individual on the board went through rigorous rounds of assessment to prove their worth. After which they invested a large part of their time researching and compiling the most relevant data for our readers. Conferences and sessions were held from time to time between the editorial board and the contributing authors to present the data in the most comprehensible form. The editorial team has worked tirelessly to provide valuable and valid information to help people across the globe.

Every chapter published in this book has been scrutinized by our experts. Their significance has been extensively debated. The topics covered herein carry significant findings which will fuel the growth of the discipline. They may even be implemented as practical applications or may be referred to as a beginning point for another development. Chapters in this book were first published by InTech; hereby published with permission under the Creative Commons Attribution License or equivalent.

The editorial board has been involved in producing this book since its inception. They have spent rigorous hours researching and exploring the diverse topics which have resulted in the successful publishing of this book. They have passed on their knowledge of decades through this book. To expedite this challenging task, the publisher supported the team at every step. A small team of assistant editors was also appointed to further simplify the editing procedure and attain best results for the readers.

Our editorial team has been hand-picked from every corner of the world. Their multi-ethnicity adds dynamic inputs to the discussions which result in innovative outcomes. These outcomes are then further discussed with the researchers and contributors who give their valuable feedback and opinion regarding the same. The feedback is then collaborated with the researches and they are edited in a comprehensive manner to aid the understanding of the subject.

Apart from the editorial board, the designing team has also invested a significant amount of their time in understanding the subject and creating the most relevant covers. They scrutinized every image to scout for the most suitable representation of the subject and create an appropriate cover for the book.

The publishing team has been involved in this book since its early stages. They were actively engaged in every process, be it collecting the data, connecting with the contributors or procuring relevant information. The team has been an ardent support to the editorial, designing and production team. Their endless efforts to recruit the best for this project, has resulted in the accomplishment of this book. They are a veteran in the field of academics and their pool of knowledge is as vast as their experience in printing. Their expertise and guidance has proved useful at every step. Their uncompromising quality standards have made this book an exceptional effort. Their encouragement from time to time has been an inspiration for everyone.

The publisher and the editorial board hope that this book will prove to be a valuable piece of knowledge for researchers, students, practitioners and scholars across the globe.

List of Contributors

Ugur Murat Leloglu and Barış Gençay
TUBITAK Space Technologies Research Institute, Turkey

Joseph A. Nuth III
Astrochemistry Laboratory, Code 691 NASA's Goddard Space Flight Center, Greenbelt, USA

Frans J. M. Rietmeijer
Dept. of Earth and Planetary Sciences, MSC03-2040, University of New Mexico, Albuquerque, USA

Cassandra L. Marnocha
University of Wisconsin at Green Bay, Green Bay, USA

Risto Pirjola
Finnish Meteorological Institute, Finland
Natural Resources Canada, Canada

Wlodarczyk Ireneusz
Chorzow Astronomical Observatory; Rozdrazew Astronomical Observatory, Poland

Alexandre C. M. Correia
University of Aveiro, Portugal

Bijay Sharma
Electronics and Communication Department, National Institute of Technology, Patna, India

Herman J. Mosquera Cuesta
Departamento de Física, Centro de Ciências Exatas e Tecnológicas (CCET), Universidade Estadual Vale do Acaraú, Sobral, Ceará, Brazil
Instituto de Cosmologia, Relatividade e Astrofísica (ICRA-BR), Centro Brasileiro de Pesquisas Físicas, Urca Rio de Janeiro, RJ, Brazil
International Center for Relativistic Astrophysics Network (ICRANet), Pescara, Italy
International Institute for Theoretical Physics and High Mathematics Einstein-Galilei, PRATO, Italy

Gaetano Lambiase
Dipartimento di Fisica "E. R. Caianiello", Universitá di Salerno, Fisciano (Sa), Italy
INFN, Sezione di Napoli, Italy

Hao Liu
Key Laboratory of Particle Astrophysics, Institute of High Energy Physics,
Chinese Academy of Sciences, China

Ti-Pei Li
Department of Physics and Center for Astrophysics, Tsinghua University, Key Laboratory of
Particle Astrophysics, Institute of High Energy Physics, Chinese Academy of Sciences, China

Printed in the USA
CPSIA information can be obtained
at www.ICGtesting.com
JSHW011341221024
72173JS00003B/187

9 781632 395733